I0502841

The Development and Application of a Decision Support System for Land Management in the Lake Tahoe Basin— The Land Use Simulation Model

By William M. Forney, I. Benson Oldham, and Neil Crescenti

Scientific Investigations Report 2012–5229

U.S. Department of the Interior
U.S. Geological Survey

U.S. Department of the Interior
KEN SALAZAR, Secretary

U.S. Geological Survey
Suzette M. Kimball, Acting Director

U.S. Geological Survey, Reston, Virginia: 2013

For more information on the USGS—the Federal source for science about the Earth, its natural and living resources, natural hazards, and the environment, visit http://www.usgs.gov or call 1-888-ASK-USGS

For an overview of USGS information products, including maps, imagery, and publications, visit http://www.usgs.gov/pubprod

Suggested citation:
Forney, W.M., Oldham, I.B., Crescenti, N. 2013, The development and application of a decision support system for land management in the Lake Tahoe Basin—The Land Use Simulation Model: U.S. Geological Survey Scientific Investigations Report 2012–5229, 52 p.

Contents

Abstract...1

Background...1

    Lake Tahoe Basin Natural and Anthropogenic Setting 2

        Fire in the Wildland-Urban Interface...5

        Land-Use Change...5

        Urban Biodiversity..6

        Water Quality and Impact of Urbanization 7

        Land Capability...8

            Bailey Land Capability System 8

            Individual Parcel Evaluation System 8

    Lake Tahoe Land Management 9

        Land-Acquisition Programs...10

        Tahoe Regional Planning Agency 10

            Allocations and Commodities 13

            Redevelopment, Mixed Use and Compact, Form-Based Design...........................15

        Other Agency Jurisdictions 15

        Pathway 2007 15

        Five Counties and One City 16

Materials, Data Sources, and Conversions...16

        Data Assessment..19

Methods, Model Design, and Logic 20

    Design Criteria for the Decision Support System..20

        Collaborator Communication and Feedback..20

            Determining the "What" and Amount of Allocation Pool(s)...........................20

            Determining the "Where" to Develop and Retire...21

            Additional Considerations 21

    Modeling Theory and Approach 21

        Decision Rules..24

        Development Intention..25

        Retirement Intention...26

        Logic and Functional Considerations 26

        Assumptions 28

Results 29

    Reasonability Analysis and Model Testing...29

    Select Model Output: Tables and Maps...29

    Discussion...35

    Limitations...45

    Suggested Next Steps 46

        Data Management for TRPA 46

        Maintenance and Updates..46

        Additional Research Avenues 46

Summary and Conclusion..47

Acknowledgments..48

References...49

Figures

1. Map showing latitude and longitude and various vegetation types in the Lake Tahoe Basin study area..3

2. Map showing an overview of Lake Tahoe Basin and inputs to the Land Use Simulation Model such as parcels, Tahoe Regional Planning Agency plan area statements, the California-Nevada border, county and city boundaries, and U.S. Forest Service lands..11

3. Conceptual schematic for the Land Use Simulation Model 17

4. Coefficients of variation results for each commodity type and increasing iteration used in the Land Use Simulation Model...30

5. Map showing output of the Land Use Simulation Model for the Lake Tahoe Basin of vacant parcels transitioning to open space or retired parcel allocations...38

6. Map showing output of the Land Use Simulation Model for the Lake Tahoe Basin of vacant parcels transitioning to single-family dwelling allocations...39

7. Map showing output of the Land Use Simulation Model for the Lake Tahoe Basin of vacant parcels transitioning to multifamily dwelling (MFD), 2 to 4 unit allocations...40

8. Map showing output of the Land Use Simulation Model for the Lake Tahoe Basin of vacant parcels transitioning to multifamily dwelling (MFD), 5 to 10 unit allocations...41

9. Map showing output of the Land Use Simulation Model for the Lake Tahoe Basin of vacant parcels transitioning to multifamily dwelling (MFD), 10+ unit allocations. 42

10. Map showing output of the Land Use Simulation Model for the Lake Tahoe Basin of vacant parcels transitioning to tourist accommodation allocations. 43

11. Map showing output of the Land Use Simulation Model for the Lake Tahoe Basin of vacant parcels transitioning to commercial allocations. 44

Tables

1. Lake Tahoe existing vegetation types 4

2. Total Maximum Daily Load pollutant loading estimates for Lake Tahoe 8

3. Bailey Land Capability System (BLCS) table with categories used for determining BLCS class and Individual Parcel Evaluation System cross-walk to fill in data gaps between the two systems. 9

4. Recommended land coverage by Bailey Land Capability System capability class for residential parcels in the Lake Tahoe Basin............................9

5. Criteria used to score parcels for the Individual Parcel Evaluation System in the Lake Tahoe Basin. 10

6. Historical activities of land acquisition programs in the Lake Tahoe Basin—parcels acquired by agency per year, 1982 to 2007, and summary statistics by program...12

7. Comparison between the 1987 regional plan and the proposed transect zoning map of the Tahoe Regional Planning Agency. 13

8. Historical residential allocations given by the Tahoe Regional
   Planning Agency that have been used and rolled over by county and
   city jurisdiction for the Lake Tahoe Basin. 14

9. Total historical residential allocations given by the Tahoe Regional
   Planning Agency that have been used and rolled over for the entire
   Lake Tahoe Basin on an annual basis...14

10. Description of geospatial datasets used in the Land Use Simulation
    Model and this report. 18

11. Characterization of the parcel dataset for the Lake Tahoe Basin by geographic unit............19

12. User variables, their descriptions, and the default values for the
    Land Use Simulation Model...23

13. Residential allocation performance table used in the Land Use Simulation Model. 25

14. Organizational framework for the Land Use Simulation Model..27

15. Fixed variables and decision rules that persist in the background of the Land Use
    Simulation Model and can be changed by select users. 28

16. Coefficients of variation by allocation type used in the Land Use Simulation Model............31

17. Annual number of transitions by land-use type as produced
    by the Land Use Simulation Model...32

18. Annual pool size of remaining vacant parcels available for transition as
    produced by the Land Use Simulation Model. 33

19. Summary of vacant parcels entering into a particular commodity with
    three levels of probability as produced by the Land Use Simulation
    Model for the Lake Tahoe Basin. 34

20. Land-cover impacts in the Lake Tahoe Basin resulting from the Land Use Simulation
    Model's default scenario's land-use transitions by commodity type. 36

Acronyms and Abbreviations

APN	Assessor's parcel number
BLCS	Bailey Land Capability System
BMP	Best management practice
CEP	Community Enhancement Program
CONCEPTS	Conservational Channel Evolution and Pollutant Transport System
CTC	California Tahoe Conservancy
CWPP	Community Wildfire Protection Plans
DSS	Decision support system
EIP	Environmental Improvement Program
EMC	Event mean concentration
GIS	Geographic information systems
GUI	Graphical user interface
IPES	Individual Parcel Evaluation System
LTIMP	Lake Tahoe Interagency Monitoring Program
LRWQCB	Lahontan Regional Water Quality Control Board, California
LTBMU	Lake Tahoe Basin Management Unit
LUSM	Land Use Simulation Model
MFD	Multifamily dwelling
NDEP	Nevada Division of Environmental Protection
NDSL	Nevada Division of State Lands
NHD	National Hydrography Dataset
NPAA	No-Project Alternative Analysis
NRCS	Natural Resources Conservation Service
OFR	Open-File Report
PAS	Plan Area Statement
SEZ	Stream Environment Zone
SFD	Single family dwelling
SNPLMA	Southern Nevada Public Land Management Act
TAU	Tourist accommodation unit
TERC	Tahoe Environmental Research Center, University of California at Davis
TDR	Transfer of development rights
TDSS	Tahoe Decision Support System
TLOS	Transit Level of Service
TEGIS	Tahoe Environmental Geographic Information System
TMDL	Total Maximum Daily Load
TRG	Tahoe Research Group, University of California at Davis
TRPA	Tahoe Regional Planning Agency
USDA	U.S. Department of Agriculture
USFS	U.S. Forest Service
USGS	U.S. Geological Survey
WGSC	USGS Western Geographic Science Center
WUI	Wildland-urban interface

The Development and Application of a Decision Support System for Land Management in the Lake Tahoe Basin— The Land Use Simulation Model

By William M. Forney[1], I. Benson Oldham[2], and Neil Crescenti[3]

Abstract

This report describes and applies the Land Use Simulation Model (LUSM), the final modeling product for the long-term decision support project funded by the Southern Nevada Public Land Management Act and developed by the U.S. Geological Survey's Western Geographic Science Center for the Lake Tahoe Basin. Within the context of the natural-resource management and anthropogenic issues of the basin and in an effort to advance land-use and land-cover change science, this report addresses the problem of developing the LUSM as a decision support system. It includes consideration of land-use modeling theory, fire modeling and disturbance in the wildland-urban interface, historical land-use change and its relation to active land management, hydrologic modeling and the impact of urbanization as related to the Lahontan Regional Water Quality Control Board's recently developed Total Maximum Daily Load report for the basin, and biodiversity in urbanizing areas. The LUSM strives to inform land-management decisions in a complex regulatory environment by simulating parcel-based, land-use transitions with a stochastic, spatially constrained, agent-based model. The tool is intended to be useful for multiple purposes, including the multiagency Pathway 2007 regional planning effort, the Tahoe Regional Planning Agency (TRPA) Regional Plan Update, and complementary research endeavors and natural-resource-management efforts. The LUSM is an Internet-based, scenario-generation decision support tool for allocating retired and developed parcels over the next 20 years. Because USGS staff worked closely with TRPA staff and their "Code of Ordinances"[4] and analyzed datasets of historical management and land-use practices, this report accomplishes the task of providing reasonable default values for a baseline scenario that can be used in the LUSM. One result from the baseline scenario for the model suggests that all vacant parcels could be allocated within 12 years. Results also include: assessment of model functionality, brief descriptions of the 7 basic output tables, assessment of the rate of change in land-use allocation pools over time, locations and amounts of the spatially explicit probabilities of land-use transitions by real estate commodity, and analysis of the state change from today's existing land cover to potential land uses in the future. Assumptions and limitations of the model are presented. This report concludes with suggested next steps to support the continued utility of the LUSM and additional research avenues.

Background

Under the regulatory framework of the Federal Clean Water Act, the Environmental Protection Agency (EPA) has designated Lake Tahoe as Outstanding National Resource Water (EPA, 2012). It considers noncontact recreation (in other words, aesthetic and visual enjoyment of the clarity of the Lake) as a primary beneficial use. The Nevada Department of Environmental Protection (NDEP) has designated the lake as a water of "extraordinary ecological or aesthetic value" (NDEP, 2012). The lake has been designated as "impaired" by the EPA because of lack of achievement of numerical standards for water quality related to input of nitrogen, phosphorous, and sediment, and thus placed on the Federal Clean Water Act Section 303(d) list (LRWQCB and NDEP, 2009). As part of the listing process, the reasons determined for its impairment are an excess supply of nutrients (nitrogen and phosphorous) and fine sediment particles to the lake. Alteration of the landscape and other manmade disturbances have been shown to be important factors affecting mass transport (supply and loading) of principal plant nutrients (nitrogen and phosphorous) and sediment to the lake (LRWQCB and NDEP, 2009). This problem has been developing in the Lake Tahoe Basin over the course of decades.

The U.S. Geological Survey's (USGS) Western Geographic Science Center's (WGSC) involvement in the Lake Tahoe Basin has been lengthy. The latest USGS project effort began in 2004 with the receipt of a $500,000 a year grant from the Southern Nevada Public Land Management Act (SNPLMA), titled the Tahoe Decision Support System (TDSS). From the words of the TDSS grant proposal, the intent of the system was to "provide a GIS-based land planning tool considering socioeconomic and environmental impacts of agency-identified controls aimed at attaining environmental standards . . . . The TDSS aims to serve as a bridge between some of the many relevant efforts in the basin—the stakeholder analyses and identification of critical controls that the agencies have undertaken, the Adaptive Management Framework effort, and appropriate scientific modeling efforts."

The TDSS project has involved USGS staff, Tahoe Regional Planning Agency (TRPA) staff, University of California at Davis Tahoe Research Group[5] (TRG) staff, Lahontan Regional Water Quality Control Board (LRWQCB) staff, and others. One of the first major efforts by USGS staff was the writing of an unpublished report submitted to the TRPA in 2004 called the No-Project Alternative Analysis (NPAA) (Duffie, and others, 2004). To assist with informing the Pathway 2007 effort (described below), the report "presents the likely effects on . . . interim indicators of a 'No-project Alternative,' in which management, demographic and climate trends in the Lake Tahoe Basin continue in their current presumed courses over the next 20 years. This No-Project Alternative Analysis is intended as a dry run of the exercises of characterizing management controls, specifying reasonable assumptions about the future, and creating a system representation that allows that alternative future to be mapped into its likely effect on indicators of basin health." In early 2005, a second product of the USGS was an analysis of the NPAA report in the Journal of Nevada Water Resources Association (Halsing and others, 2005). The article included identification of determinants of threshold indicator attainment, projections, knowledge gaps and considerations for the development of the TDSS. The primary focus of the model was to simulate land use and preservation, which would then feed into (or be related to) models of population, transportation, air quality, socioeconomics, watershed dynamics, climate change, wildlife, lake clarity, and other factors. From the research done for the unpublished NPAA report and the Halsing and others (2005) article, the groundwork was developed for a Land Use Scenario Generation Model to enable scenario generation (of land-use and population components) and system visualization for natural-resource management purposes.

Over time and the course of the SNPLMA grant, the TDSS project team became more focused on informing LRWQCB Total Maximum Daily Load (TMDL) watershed modeling effort. In fact, the TDSS was used to generate sediment- and nutrient-loading estimates from event mean concentrations (EMC) on pervious and impervious surfaces, which were aggregated from a parcel to a subwatershed level and then provided to Tetra Tech's (2007) TMDL model in a "worst case" scenario (that is, the maximum possible loading and the largest negative environmental impact to the natural system)

(Halsing, 2006). Reworked as the Tahoe Land Use Change Model (Hessenflow and Halsing, 2006), the back-end of the model was extended to meet the preliminary needs of TRPA and a front-end graphical user interface (GUI) was developed in an attempt to suit the needs of both LRWQCB and TRPA. The Land Use Change Model's platform evolved from a desktop-based geographic information systems (GIS) platform and Python code language (back end that is not visible to the user) to an Internet-based Servoy© platform and PostGreSQL and Java© coding languages (front end that is presented to the user). Early in 2008, the USGS and TRPA scoped out the objectives of the final efforts for the TDSS as the Land Use Simulation Model (LUSM). With an on-line decision support tool in mind, the objectives were to integrate the back end and front end of the model, incorporate updated data, disaggregate its output to the parcel level, consider the possibility and feasibility of model extensions, finish out the remaining contracts for LRWQCB and TRPA, and administration of grants from SNPLMA.

Lake Tahoe Basin Natural and Anthropogenic Setting

Lake Tahoe is more than 35 kilometers (km) long and 19 km wide (fig. 1). It lies at the eastern edge of the Sierra Nevada and is bordered by the Crystal Range on the west and the Carson Range on the east. It has almost 116 km of shoreline, and the surface area of the lake covers approximately 500 square kilometers (km^2). At a maximum depth of 501 meters (m), Lake Tahoe is the 2nd deepest lake in the United States and the 11th deepest lake in the world (Tahoe Environmental Research Center, 2009). The hydraulic residence time is 650 years, and the lake remains ice-free year round (Roberts and Reuter, 2007). Geologically, it is a graben lake, which is formed by the subsidence of the fault block underlying Lake Tahoe. The basin surrounding the lake lies between the elevations of 1,860 and 3,317 m and has a total watershed area of approximately 1,300 km^2, of which the lake accounts for approximately 38 percent. As shown in figure 1, the landscape is covered by various vegetation types (Dobrowski and others, 2006; Raumann and Cablk, 2008), including wetlands, alpine meadows, shrublands (such as montane chaparral, alpine scrub, and Great Basin sagebrush plant communities), mixed-conifer stands (such as ponderosa, sugar, and lodgepole pines (*Pinus ponderosa*, *Pinus lambertiana*, and *Pinus contorta*), white and red fir (*Abies concolor* and *Abies magnifica*), and incense cedar (*Calocedrus decurrens*)) and riparian and upland deciduous stands (such as quaking aspen (*Populus tremuloides*), mountain alder (*Alnus incana*), black cottonwood (*Populus trichocarpa*) and various willows (*Salix* spp.)), and granitic- and andesitic-derived soils. The basin is about 85 percent National Forest. Table 1 characterizes the vegetation cover types, with the predominant ones being Jeffery pine (*Pinus jeffreyi*), white fir, red fir, and Great Basin sagebrush (*Artemisia tridentate*).

---

[5]Now called the Tahoe Environmental Research Center (TERC).

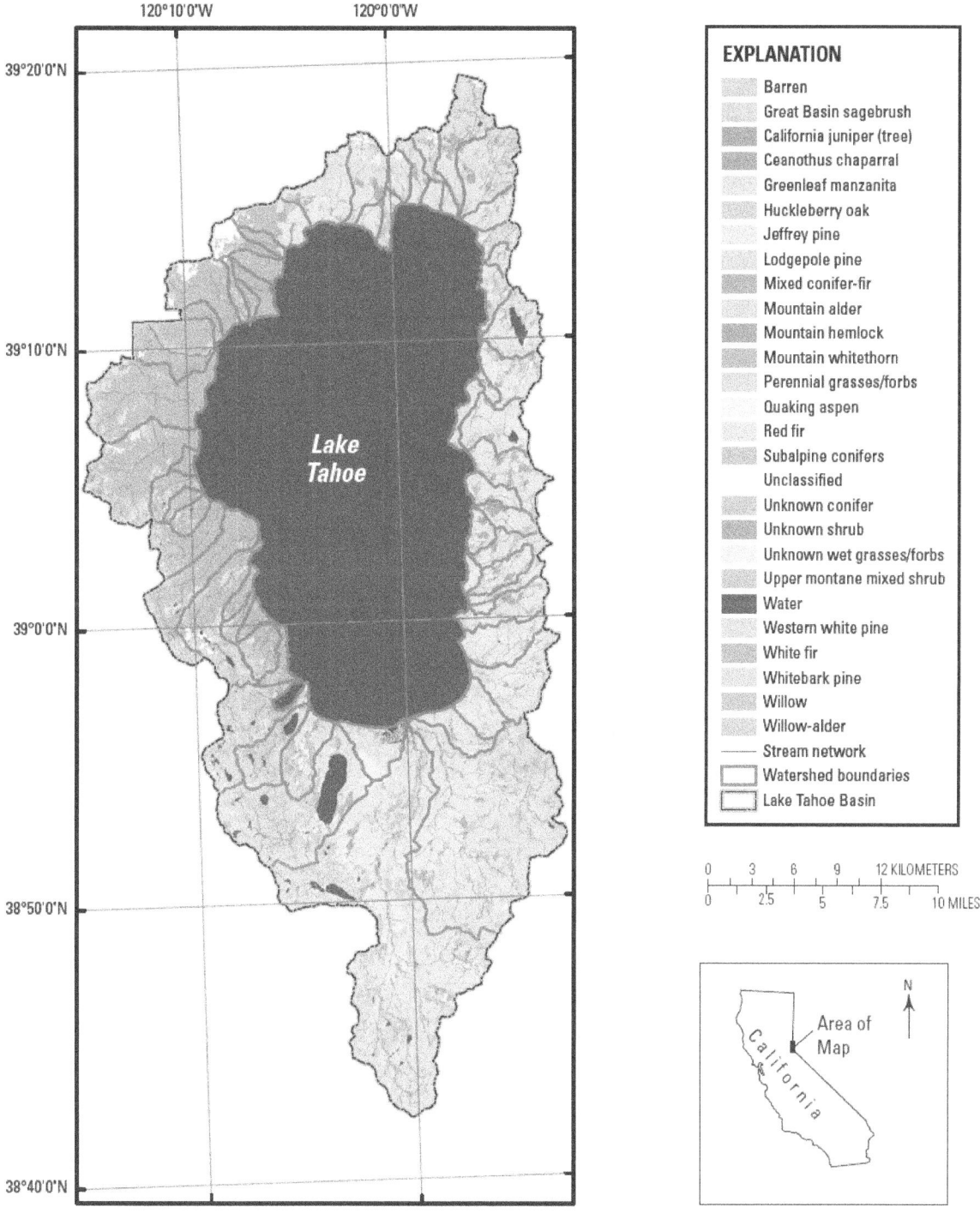

Figure 1. Map showing latitude and longitude and various vegetation types in the Lake Tahoe Basin study area. Primary types are ponderosa, sugar, and lodgepole pines (*Pinus ponderosa, Pinus lambertiana,* and *Pinus contorta*); white and red fir (*Abies concolor* and *Abies magnifica*); incense cedar (*Calocedrus* sp.); riparian and upland deciduous stands, such as quaking aspen (*Populus tremuloides*), mountain alder (*Alnus incana*), black cottonwood (*Populus trichocarpa*), various willows (*Salix* spp.), Jeffery pine (*Pinus jeffreyi*), and Great Basin sagebrush (*Artemisia tridentate*); California juniper (*Juniperus californica*); Ceanothus chaparral (*Ceanothus* sp.); greenleaf manzanita (*Arctostaphylos patula*); huckleberry oak (*Quercus vacciniifolia*); mountain hemlock (*Tsuga mertensiana*); mountain whitethorn (*Ceanothus cordulatus*); western white pine (*Pinus monticola*); and whitebark pine (*Pinus albicaulis*). Base map sources: Dobrowski and others (2006); table 10.

Table 1. Lake Tahoe existing vegetation types and statistical characterization of their distribution and size (from Dobrowski and others, 2006).

[ha, hectares, ponderosa, sugar, and lodgepole pines (*Pinus ponderosa, Pinus lambertiana,* and *Pinus contorta*), white and red fir (*Abies concolor* and *Abies magnifica*), incense cedar (*Calocedrus* sp.), quaking aspen (*Populus tremuloides*), mountain alder (*Alnus incana*), black cottonwood (*Populus trichocarpa*), various willows (*Salix* spp.), Jeffery pine (*Pinus jeffreyi*), Great Basin sagebrush (*Artemisia tridentate*), California juniper (*Juniperus californica*), Ceanothus chaparral (*Ceanothus* sp.), greenleaf manzanita (*Arctostaphylos patula*), huckleberry oak (*Quercus vacciniifolia*), mountain hemlock (*Tsuga mertensiana*), mountain whitethorn (*Ceanothus cordulatus*), western white pine (*Pinus monticola*), and whitebark pine (*Pinus albicaulis*)]

Land-cover type	Number of patches	Total area, ha	Average area, ha	Standard deviation, ha
Jeffrey pine	1,889	15,679.30	8.3	8.6
White fir	1,977	14,926.70	7.5	8
Red fir	1,875	14,537.00	7.7	8.2
Basin sagebrush	4,940	10,215.20	2.1	1.9
Huckleberry oak	1,740	5,250.50	3	2.5
Upper montane mixed Shrub	1,515	4,203.60	2.8	2.7
Subalpine conifers	583	2,890.50	5	5.7
Mixed conifer-fir	220	2,082.70	9.5	10.1
Lodgepole pine	379	1,732.30	4.6	4.6
Greenleaf manzanita	673	1,655.30	2.5	2.1
Perennial grasses/forbs	566	1,411.50	2.5	1.9
Whitebark pine	392	1,353.90	3.4	2.5
Willow	941	1,324.00	1.4	1.4
Water	180	1,313.80	7.3	11.9
Quaking aspen	364	1,205.70	3.3	4.2
Barren	508	871.4	1.7	1.7
Unclassified	1,569	860	0.5	1
Unknown conifer	72	550.6	7.6	8.3
Ceanothus chaparral	90	340.7	3.8	2.4
Western white pine	100	306.2	3.1	2.3
Mountain whitethorn	31	145.1	4.7	3.6
Mountain hemlock	27	140.8	5.2	4.4
Unknown wet grasses/forbs	37	113.5	3.1	3
Unknown shrub	31	62.9	2	2.7
Mountain alder	50	53.4	1.1	0.8
California juniper (tree)	7	26.3	3.7	2.9
Willow-alder	12	11.8	1	0.9
Western white pine	100	306.2	3.1	2.3
Mountain whitethorn	31	145.1	4.7	3.6
Mountain hemlock	27	140.8	5.2	4.4
Unknown wet grasses/forbs	37	113.5	3.1	3
Unknown shrub	31	62.9	2	2.7
Mountain alder	50	53.4	1.1	0.8
California juniper (tree)	7	26.3	3.7	2.9
Willow-alder	12	11.8	1	0.9

Annual precipitation ranges from more than 76 centimeters (cm) on the east side of the Lake Tahoe Basin to 178 cm on the west side of the basin (Rowe and others, 2002), and most is received during the winter months. The average annual total precipitation is 83.2 cm, of which 6.2 cm occurs in the summer months. In the winter, the average air temperature is −0.6 degrees Celsius (°C) and the average daily minimum temperature is -6.4°C. In summer, the average air temperature is 14.7°C and the average daily maximum temperature is 23.6°C (Loftis, 2007). Sixty-three watersheds and 44 intervening areas contribute runoff to Lake Tahoe (Jassby and others, 1994), and only one outlet—the Lower Truckee River—drains the lake from Tahoe City to the east out of the mountains. The Truckee River terminates at land-locked Pyramid Lake in Nevada.

The following sections address topics that are important to natural-resource managers in the basin and influence land-management decisions. In addition to relevant literature review and discussion of management activities related to the development and application of the LUSM, the section topics include fire in the wildland-urban interface (WUI), historical land-use change, urban biodiversity, water quality and the impact of urbanization, the basin's land capability classification systems, and relevant regulatory institutions.

Fire in the Wildland-Urban Interface

In the western United States, wildfire is a great concern in natural-resource management and land-use planning. In 2008 in California, more than 5,812 wildfires were ignited and more than 542,210 hectares (ha) were burned. In 2009 in California, more than 9,150 wildfires were ignited and more than 164,130 ha were burned (National Interagency Fire Center, 2010). On June 24, 2007, the Angora Fire ignited in South Lake Tahoe near Meyers, California, and burned approximately 1,290 ha and 250 homes. For decades, the common practice was to suppress forest fires. In the Lake Tahoe Basin, this practice began in the 1920s (Raumann and Cablk, 2008). Considering a nonsuppression fire disturbance regime, at least three to five natural fire cycles (Manley and others, 2000) and at most eight natural fire cycles (Steve Holl Consulting and Wildland Rx, 2007) would have occurred since the 1920s, which would have thinned stands and removed hazardous fuels. Fire and fuel behavior models have indicated that the most severe fires would likely have occurred in lower elevation pine and mixed conifer forests (Steve Holl Consulting and Wildland Rx, 2007).

The Office of the State of California's Fire Marshall mandated that building codes were to be updated on September 20, 2007. Chapter 7a of the Title 24 code requirements regulates new buildings in any fire hazard severity zone (California Department of Forestry and Fire Protection, 2008). The regulation prescribes standards for such items as exterior wall siding and sheathing, exterior windows and doors, attic ventilation and eave protection, and decking surfaces and floor protection. Also, in 2004, local jurisdictions in the basin completed Community Wildfire Protection Plans (CWPP) to

identify and reduce fuels over the following 10-year period. The CWPPs include strategies such as creating neighborhood fuel reduction zones, defensible space on properties, and fire-wise landscaping; improving signage on streets, houses, and properties; inspecting annually for code and regulation compliance; installing fire-safe roofing; restoring fire disturbance in ecosystems; finding ongoing funding for compliance efforts; improving use of remnant biomass; and improving community education and cooperation and coordination (Citygate Associates, LLC, 2004). The U.S. Forest Service (USFS) is working to improve the health of forests and to minimize the hazard faced in the WUI from overstocked forests as evidenced by more than 5,250 ha of fuel hazard reduction between 2000 and 2006 (Steve Holl Consulting and Wildland Rx, 2007). Fuels, such as downed logs, coarse wood debris, underbrush, and snags, provide essential habitat for various wildlife species in urban forest lots but were found to decrease in remnant forests and the surrounding landscape as a result of development (Manley and others, 2007), thereby potentially putting alternate natural-resource objectives for a given parcel at odds with each other.

Land-Use Change

Since the 1850s, the Lake Tahoe Basin has been shaped by anthropogenic factors and land-use processes. From the 1850s to the 1920s, the basin was extensively logged for building materials and to support various mining activities (Richards, 1999). Transportation infrastructure went into place in the 1930s, and since the 1950s the basin has been increasingly urbanized. With Squaw Valley hosting the winter Olympics in 1960, the visibility of the area as a tourist and recreational destination greatly increased, and the population has increased fivefold (Tahoe Regional Planning Agency, 1996). As part of the Sierra Nevada Ecosystem Project, McGurk and Davis (1996) used temporal land use and road mapping to assess 50 years of the hydrologic effects of land management in two creeks in the basin. They found that disturbances to natural systems were caused primarily by logging, road building, and residential development. Legacy sites such as these where the impacts from past disturbances still persist offer great potential to efficiently reduce loading from forested areas (LRWQCB and NDEP, 2008). Having done a multitemporal change detection analysis, Raumann and Cablk (2008) found that between 1940 and 2002 the most significant land conversion was to developed lands, with a corresponding decrease in forest, wetland, and shrubland cover types. Furthermore, they found the predominant causes of landscape change to include regional population growth, tourism demands, timber harvest for local use, fire suppression, bark beetle attack, and fuels reduction activities. They also found that the highest rate of development in the South Lake Tahoe Basin study area between 1940 and 1969 was residential commodities (such as single-family dwellings), with the second highest rate of development being commercial commodities (such as mixed-use shopping districts) (Raumann and Cablk, 2008). Population and development rates began to taper off in the mid-1980s, presumably because of constraints on land availability and increased regulatory activity. This observation is supported by the fact that only 0.2 ha of wetlands were lost between 1987 and 2002 (Raumann and Cablk, 2008). Increased constraints and regulatory activity included (1) the inception of TRPA in 1969, (2) the establishment of Stream Environment Zones (SEZ) as an important biophysical landscape unit and the development restrictions placed on them in 1981, and (3) the adoption of the TRPA Regional Plan in 1987. The residual effects of historical land-use change persist into the present; thus, a historical perspective included in current analyses of landscape structure may contribute to land-management systems that are more relevant to the dynamic systems they are intended to manage (With, 2007).

As of 1996, more than 8 million people lived within a few hours of the Lake Tahoe Basin, including those in the major metropolitan areas of San Francisco and Sacramento, California, and Reno-Carson City, Nevada. This proximity provides a steady flow of tourists and recreationists to the basin to enjoy activities such as hiking, skiing, biking, water sports, sightseeing, shopping, and casino gaming (Raumann and Cablk, 2008). Twenty-three million people visit the basin annually, and it supports a permanent population of approximately 55,000, with weekend populations swelling to more than 200,000 (Elliot-Fisk and others, 1997). The population of the basin is heavily influenced by overnight, transient, and seasonal visitors (Duffie and others, 2004). More recent estimates of the permanent population based on the 2000 U.S. Census figures have been closer to 62,000 or 63,000 (Duffie and others, 2004). Retrospective analyses have projected an annual growth rate of 1.8 percent, which—if assumed to be constant—would bring the total number of residents by 2027 to approximately 90,000. Recent estimates of land ownership indicate that private ownership in the Lake Tahoe Basin is about 12 percent (Elliot-Fisk and others, 1997). Given the nature, growth, and demand for resource use and the preferences of its human inhabitants related to settlement patterns and other related factors, a fundamental question arises—How and where can these populations be served in a manner that sustains the natural resources and ecosystem services that draw people to the Lake Tahoe region?

Linking urban development and historical land-use change to current human usage patterns is important to understanding the environmental-impact and natural-resource ramifications of the land-use evolution of a landscape. In a study in Teton Valley, Idaho, which is also a year-round tourist- and recreational-based economy like the Lake Tahoe Basin, Peterson and others (2008) found interesting relations among household location choices and their influence on biodiversity conservation. The study found that the lower educated and least environmentally inclined newcomers settled in previously established residential areas, whereas the higher educated and most environmentally inclined newcomers settled in sensitive natural areas (that is, areas such as riparian zones and wildlife corridors) and

undertook new construction. This latter population had greater environmental impacts because of the higher likelihood of small-sized households (less than 3 people per household) in larger homes. Furthermore, and counter intuitively, longer residency in the natural areas was predicted to create less environmentally oriented attitudes and a lower care for nature. The authors (Peterson and others, 2008) called for, "explicit consideration of household location decisions on resource use and biodiversity conservation, and development of ways to experience pristine environments besides building houses on them." In the Lake Tahoe Basin, 40 percent to 54 percent (Duffie and others, 2004) of the total housing stock is dedicated to seasonal or second home occupancy. Although these anthropogenic phenomena may be directly or indirectly related to the well-being of Lake Tahoe's ecology and influence different interrelated drivers of change (Halsing and others, 2005), they frame issues facing land-management agencies as they work to balance prosperity for both the natural and built environments. Understanding the phenomenon of population growth in the physically constrained Lake Tahoe landscape and the concomitant environmental impacts and efforts to mitigate these impacts through informed decision making is the focus of the LUSM. The LUSM uses the parcel level as the unit of analysis for various stakeholders, institutions, and agents of change.

Urban Biodiversity

Land-use conversion and development can create conduits for the introduction of exotic plant species (Dramstad and others, 1996), which can out-compete and displace native vegetation, resulting in the loss of habitats and the wildlife they support. In the Lake Tahoe Basin, Manley and others (2007) linked indirect resource use by humans to impacts on wildlife, such as birds, mammals, ants, and plants. Manley and others (2007) found useful indicators of ecological sensitivity to urbanization to be forest structure; snag preservation and understory retention; bird species richness and dominance; raccoon (*Procyon lotor*) and coyote (*Canis latrans*) behavior (species well adapted to developed areas); martens (*Martes americana*) spotted skunk (*Spilogale gracilis*), and bobcat (*Lynx rufus*) behavior (species averse to developed areas); and particular ant species. For most of the potential indices, key questions for additional research revolved around the landscape configurations and human management activities that would be most beneficial to the ecology, wildlife populations and their associated behaviors, and ecosystem services (which are the goods and services provided to humans from natural ecosystem resources and processes) (Manley and others, 2007). An interesting statistical analysis was that the Normalized Difference Vegetation Index (an indicator used to analyze remote-sensing measurements of live, green vegetation) and canopy cover were significant, negative indicators of development pressure. In other words, greater amounts of vegetation and canopy density around the basin correlate to lower amounts of human development and urbanization. This relation is corroborated by Raumann and Cablk's (2008) multitemporal change-detection analysis. This intuitively makes sense, as a developed parcel would be expected to have less vegetation, and more of the understory would be expected to be cleared. Manley and others (2007) also found that increased levels of development facilitate the invasion of shade-intolerant, exotic species and increased nutrient inputs to support their growth.

Water Quality and Impact of Urbanization

Alteration of the landscape and other human-caused disturbances have been shown to be important factors affecting mass transport and loading of principal plant nutrients (nitrogen and phosphorous) and sediment to Lake Tahoe. Water-quality assessment is often viewed as an integrated environmental indicator of ecosystem function and stress (Berka and others, 1995). The increase of bioavailable nutrients and sediment loading over time is suspected as a principal cause for the increase in algal growth in the lake, with a concomitant decrease in water clarity (Forney and others, 2001). Riparian ecosystems can be categorized by their nutrient dynamics and water-quality effects through existing processes in the channel and hyporheic zone (the area adjacent to and just below the channel where surface water mixes with groundwater) (Merrill, 2001). Riparian buffers can trap sediment and uptake nutrients, a function that is compromised when they are drained, altered, or encroached on (Weller and others, 1997). Forney and others (2001) hypothesized that changes in the extent of urban growth, increases in impervious surfaces, and decreases in natural vegetation have resulted in severe impacts on ecosystem health and integrity, riparian zones, and water quality over time. Coats and others (2006) extended this working hypothesis through statistical and geospatial analysis of watershed land cover characteristics, and water quality sampling from the Lake Tahoe Interagency Monitoring Program (LTIMP) and the Stormwater Monitoring Program. Coats and others (2008) concluded that impervious-surface and residential density were important factors in water-quality degradation.

The intensity of land use can be assessed by measuring the imperviousness of the surface of the landscape, with increasing concentrations of pollutants and water-quality impacts often occurring in watersheds of urbanizing areas (Arnold and Gibbons, 1996; Kauffman and Brant, 2000). Imperviousness, a measure of areas that have been built or compacted or that otherwise limit the infiltration of precipitation into the soil, is an essential measure of the potential to transport pollutants into water bodies (Forney and others, 2001), and the level of imperviousness in a watershed has been shown to influence water quality (Boothe, 1991). Coats and Goldman (2001) plumbed the value of the long-term LTIMP record and found that organic nitrogen—which is associated with soil-enriching, organic carbon versus inorganic nitrogen—accounts for more than 90 percent of the total nitrogen load in Lake Tahoe Basin streams, and the variation in annual runoff explains most of the interannual and interwatershed variability. This latter fact is related to precipitation-runoff

dynamics and how they are influenced by land use and imperviousness (Riverson and others, 2005).

In an attempt to get a solid estimate of imperviousness in the basin, Cablk and Minor (2003) used 2000 IKONOS imagery to classify impervious cover with techniques of principal component analysis and others. Despite the dense conifer canopy, they achieved an overall accuracy of 92.94 percent in a 25-km^2 test area in South Lake Tahoe. Using the same IKONOS satellite sensor with imagery from 2002, the technique was expanded to the entire Lake Tahoe Basin with similar accuracies (Minor and Cablk, 2004). Accurate estimation of impervious cover is relevant to all of the TRPA thresholds, including noise, recreation and scenic resources, wildlife and fish habitat, air and water quality, soils and vegetative cover (Cablk and Minor, 2003). TRPA had strict policies on the amount of allowable impervious coverage by land-use type and land capability class. Also, Minor and Cablk (2004) found that of the total hard impervious cover found in the entire basin, 47 percent is within 1 km of the lakeshore, and 76 percent is found within 3 km of the lakeshore.

As stated earlier, Lake Tahoe is an impaired watershed under the Federal Clean Water Act, Section 303(d). Its impairment is the result of elevated levels of nutrients (nitrogen and phosphorous) and fine sediment particle loading, with specific sources identified as urban upland runoff, nonurban upland runoff, atmospheric deposition, stream channel erosion, groundwater, and shoreline erosion (Roberts and Reuter, 2007, table 2). To restore water quality above impairment levels, the mass- or loading-based regulatory mechanism of the TMDL is required (Roberts and Reuter, 2007). The TMDL pollutant-loading estimates (table 2) helped to focus the orientation of the LUSM, particularly on urban and nonurban areas producing surface runoff to the lake.

Beyond specifically identifying sources of loading, load-reduction methods and their associated costs for a specific location are necessary to achieve tangible load reductions as required by the Clean Water Act (Veith and others, 2003). The Lake Tahoe TMDL Pollutant Reduction Opportunity Report (LRWQCB and NDEP, 2008) identified particular geographic characteristics for load-reduction opportunities including the urban and nonurban upland locations. The primary variables to categorize and stratify the targets in the urban settings were slope and impervious coverage. The largest opportunities for load reductions in almost all pollutant budgets (except for atmospheric nitrogen) are in urban[6] and groundwater settings, with forest uplands being a distant third (LRWQCB and NDEP, 2008). Implementation of urban and groundwater pollutant controls show 20-year costs ranging from $1.5 billion to $3.2 billion (LRWQCB and NDEP, 2008). 2ndNature, LLC, (2006) conducted a detailed review of existing data and reports, general best management practices (BMP) performance, and BMP engineering designs. Analyzed in terms of

---

[6]Especially in relation to fine sediment particles.

Table 2. Total Maximum Daily Load pollutant loading estimates for Lake Tahoe (from Roberts and Reuter, 2007); loading estimates of nitrogen, phosphorous, and fine sediment are categorized by source category.

[n/a, not applicable, because groundwater was assumed to not transport fine sediment particles Most of the data used to derive the estimates were collected since 2000]

Source category	Total nitrogen (metric tons/year and percent)		Total phosphorous (metric tons/year and percent)		Number of fine sediment particles ($\times 10^{18}$/year and percent)	
Urban upland runoff	63	15.9	18	39.1	348	72.3
Nonurban upland runoff	62	15.6	12	26.1	41	8.5
Atmospheric deposition (wet and dry)	218	54.9	7	15.2	75	15.6
Stream channel erosion	2	0.5	<1	<1	17	3.5
Groundwater	50	12.6	7	15.2	n/a	n/a
Shoreline erosion	2	0.5	2	4.3	1	0.2
Total (metric tons/year)	397		46		481	

inflow and outflow EMCs, the primary BMPs evaluated by 2ndNature, LLC, (2006) were dry detention basins, constructed wetlands/wet basins/meadows and mechanical treatment structures. Preliminary findings included a watershed-based approach to treatment to achieve adequate measures and acceptable levels of stormwater quality.

Land Capability

The fire dynamics in the WUI, historic land-use change, urban biodiversity, water quality, and urbanization are important to natural resource management. In many ways, anthropogenic factors and land-use development practices influence them all. Land-use management, land capability and development potential in the basin are driven predominantly by soil type and other hydrogeomorphic properties. In the basin, TRPA uses two land-capability systems—the Bailey Land Capability System (BLCS) and the Individual Parcel Evaluation System (IPES). Established in 1974, the BLCS applies to all parcels. Instituted in 1989, IPES applies to only those residential parcels that were vacant as of 1987, and consequently scored according to the system described below.

Bailey Land Capability System

During the period of January to June, 1971, Robert G. Bailey (USFS) in a cooperative study with TRPA, conducted a reconnaissance-level land-capability study of the Tahoe Basin. Bailey (1974) defined land capability, "as the level of use an area can tolerate without sustaining permanent damage through erosion and other causes." The BLCS is based on two primary factors—soil type and geomorphic setting (Bailey, 1974). Recently, the Natural Resources Conservation Service (NRCS) updated the soil survey of the basin, and the land-capability classes have been reinterpreted from that updated information. Soil type can be further characterized by depth, texture, and slope. On the basis of soil type, Bailey made

interpretations of erosion hazard, hydrologic-soil group, soil drainage, and rockiness and stoniness.

Geomorphic setting was delineated on the basis of homogeneous 2.59 km^2 areas of landform development, areas with distinctive internal structure and surface materials, and distinctive drainage patterns. Given these landscape classes and considerations, Bailey established six major groups—glaciated granitic uplands, glaciated volcanic flowlands, streamcut granitic mountain slopes, streamcut volcanic flowlands, depositional lands, and over steepened slopes. Each of these types was classified into high-, moderate-, and low-hazard lands. The four factors of soil type and the one factor of geomorphic classification were mapped, and the five maps were formalized into the BLCS (table 3). The BLCS defined and still determines the allowable coverage and impervious surface for a given parcel (table 4). Restrictions on land use result from the BLCS and riparian zones with mesic moisture regimes. Bailey assigned 88 percent of the land to the highest hazard classes, and only 6 percent to the lower hazard classes that are safer to develop (Bernknopf and others, 2003). Particular high-hazard landscape features are the SEZs, which were established in 1981 as special regulatory and management landscape units. Their classification was related to the presence of streams, high groundwater, alluvial soils, primary and secondary riparian vegetation, and geomorphic factors such as slopes, banks, and terraces (TRPA Code of Ordinances; TRPA, 2008c).

Individual Parcel Evaluation System

The IPES is a finer-resolution, more extensive system meant to improve on the coarser-resolution BLCS. In 1987, IPES assessed the land capability of the approximately 14,000 vacant residential parcels in the Lake Tahoe Basin. Over the 2-year period between 1987 and 1989, soil scientists and land managers in the basin completed extensive parcel-by-parcel field work and site-specific interpretation to create a numerical score based on physical and site characteristics. Table 5 summarizes the criteria of the system (TRPA Code of

Table 3. Bailey Land Capability System (BLCS) table with categories used for determining BLCS class and Individual Parcel Evaluation System (IPES) cross-walk to fill in data gaps between the two systems.

IPES cross-walk	BLCS class	Tolerance for use	Slope percent	Relative erosion potential	Runoff potential	Disturbance hazard
726 or greater[1]	7	Most	0–5	Slight	Low to moderately low	Low hazard lands
	6		0–16			
	5				Moderately high to high	
	4		9–30	Moderate	Low to moderately low	Moderate hazard lands
1 to 725[2]	3				Moderately high to high	
	2		30–50	High	Low to moderately low	High hazard lands
	1a		> 30		Moderately high to high	
0[2]	1b		Poor natural drainage			
	1c	Least	Fragile flora and fauna			

[1]Secondary land sensitivity parcels for the weighting of preferences for land-acquisition programs in the Land Use Simulation Model.

[2]Priority land sensitivity parcels for the weighting of preferences for land-acquisition programs in the Land Use Simulation Model.

Table 4. Recommended land coverage by Bailey Land Capability System (BLCS) capability class for parcels in the Lake Tahoe Basin.

BLCS class	Allowable percent of impervious coverage
7	30
6	30
5	25
4	20
3	5
2	1
1	1

Ordinances; TRPA, 2008c). A higher score is desirable for an owner who wishes to develop a parcel, and presently in Placer County, a minimum score of 726 is required to obtain approval for construction. In all other counties of the basin, a score of greater than 1 is required. A parcel with a score of 0 is deemed unbuildable. Bernknopf and others (2003) found that the value of an IPES score exerts a small, but significant, effect on the market value of a residential parcel in the Upper Truckee Watershed, which implies that IPES also is an effective and meaningful land-use planning and management tool. In comparison to the BLCS, the adoption of IPES applied a soil-based, hydrogeomorphic perspective to land-use classification systems, yet it recognized more criteria relevant to development. Specifically, it incorporated information about the difficulty of the parcel's access, needs for water quality improvement, complexity of the site's engineering requirements, characteristics related to ease of revegetation, and potential disturbance to SEZs. In addition, for spatial elements related to a broader geographic context, it included the overall watershed condition of where the parcel was situated and its proximity to the lake.

Lake Tahoe Land Management

Spanning the border of California and Nevada, the management of Lake Tahoe's natural and built environments is a complex web of political, regulatory, and geographic contexts (fig. 2). Although overlaps exist, each management agency has independent charges and responsibilities that influence the availability, distribution, and future of a particular parcel's land use.[7] The four primary agencies are the TRPA, the USFS, California's LRWQCB, and Nevada's NDEP. In an effort to coordinate and update their resource management plans,

[7]In this case, land use is defined to include use that is meant to support human needs (for example, single family residential) as well as ecological needs (for example, open space or preserved lands).

Table 5. Criteria used to score parcels for the Individual Parcel Evaluation System in the Lake Tahoe Basin.

Criteria	Input factors	Maximum points allocated
Relative erosion hazard	Soil samples, slope, precipitation	450
Runoff potential	Vegetative cover, infiltration	200
Access	Extent of excavation and vegetation removal	170
Stream Environment Zones	Encroachment of utilities, excavation, grading	70
Condition of watershed	Overall status of watershed	70
Ability to revegetate	Soil and site properties	50
Need for water-quality improvements	Cut and fill slopes, drainage, paved roads	50
Distance from Lake Tahoe	Aerial distance	50

the four agencies have engaged in joint, long-term planning efforts such as Pathway 2007 (Pathway, 2008). Furthermore, an alternate level of management in the basin includes county, city and regional park jurisdictions, as well as the presence Federal and State institutions, such as U.S. Army Corps of Engineers, U.S. Environmental Protection Agency and the California and Nevada Departments of Transportation. Finally, additional land-use stakeholders include various private and not-for-profit groups that add input to the public process and the preferences for management and regulatory action; these include the North Lake Tahoe Resort Association, Sierra Business Council, various Chambers of Commerce, and the League to Save Lake Tahoe. With sometimes-competing agendas and interests, these many institutions must balance aspects of the natural and built environment with the overall goal of creating prosperity for both.

The following sections discuss more specifics of the missions of these various agencies, their programs and jurisdictions, and how they relate to the allocations, decision rules, variables, and output of the LUSM. Although the charges and directives of other agencies and jurisdictions are considered, because of the nature and evolution of the SNPLMA grant and the project's administration, this project's current and primary partner is TRPA.

Land-Acquisition Programs

Land management in the Lake Tahoe Basin has a long history of public land acquisition. Starting in the 1920's, the California Department of Parks and Recreation took private lands out of circulation and put them under public ownership (Raumann and Cablk, 2008). Currently, four land acquisition programs exist to purchase and retire sensitive lands—conducted by the USFS's Urban Lot Management Program, the California Tahoe Conservancy's (CTC) Environmentally Sensitive Lands Program, the Nevada Division of State Lands' (NDSL) Nevada Tahoe Resource Team, and TRPA's Sensitive Lot Program. Although the datasets are not comprehensive in

tracking their lifespans, table 6 provides an indication of the historic activities of the four programs. These values can be analyzed to inform the default values of the LUSM. Beyond acquisition of lands or conservation easements, many of these programs work to manage, enhance, and restore their lands and to implement the Tahoe Environmental Improvement Program (EIP). State and county management of lands includes some public open space and recreational parks distributed around the basin, such as Burton Creek State Park and Sugar Pine Point State Park, which are not extensively considered in the logic and code development of the LUSM. For the future projections of the LUSM, it is assumed that additional open space or retired lands will reside with one of the four acquisition programs.

Tahoe Regional Planning Agency

TRPA is charged with, "Protecting this national treasure for the benefit of current and future generations. Our vision is to have a lake and environment that is clean, healthy and sustainable for the community and future generations" (TRPA, 2008c). TRPA is a bi-state planning agency, whose jurisdiction covers the entire 1,300-km^2 watershed, including the waters of Lake Tahoe. Created in 1969 by Congressional agreement, TRPA worked to create a regional plan, which was adopted in 1987. Since then, it has been managing the natural and anthropogenic resources of the basin, including project review and permit allocations. Currently, and in conjunction with the Pathway 2007 initiative (discussed later in this section) and the development and submittal of an Environmental Impact Statement, the TRPA staff and governing board are working to amend the 1987 regional plan (TRPA, 2008c) and the governing land-use map to redefine district boundaries, zones, densities, rates of growth, land coverage, and types of use. Using a transect method, the five most general regional plan district boundaries are to be updated and as many as ten district boundary transect zones are to be created (TRPA, 2008d). Although the boundary revisions have not been finalized as of

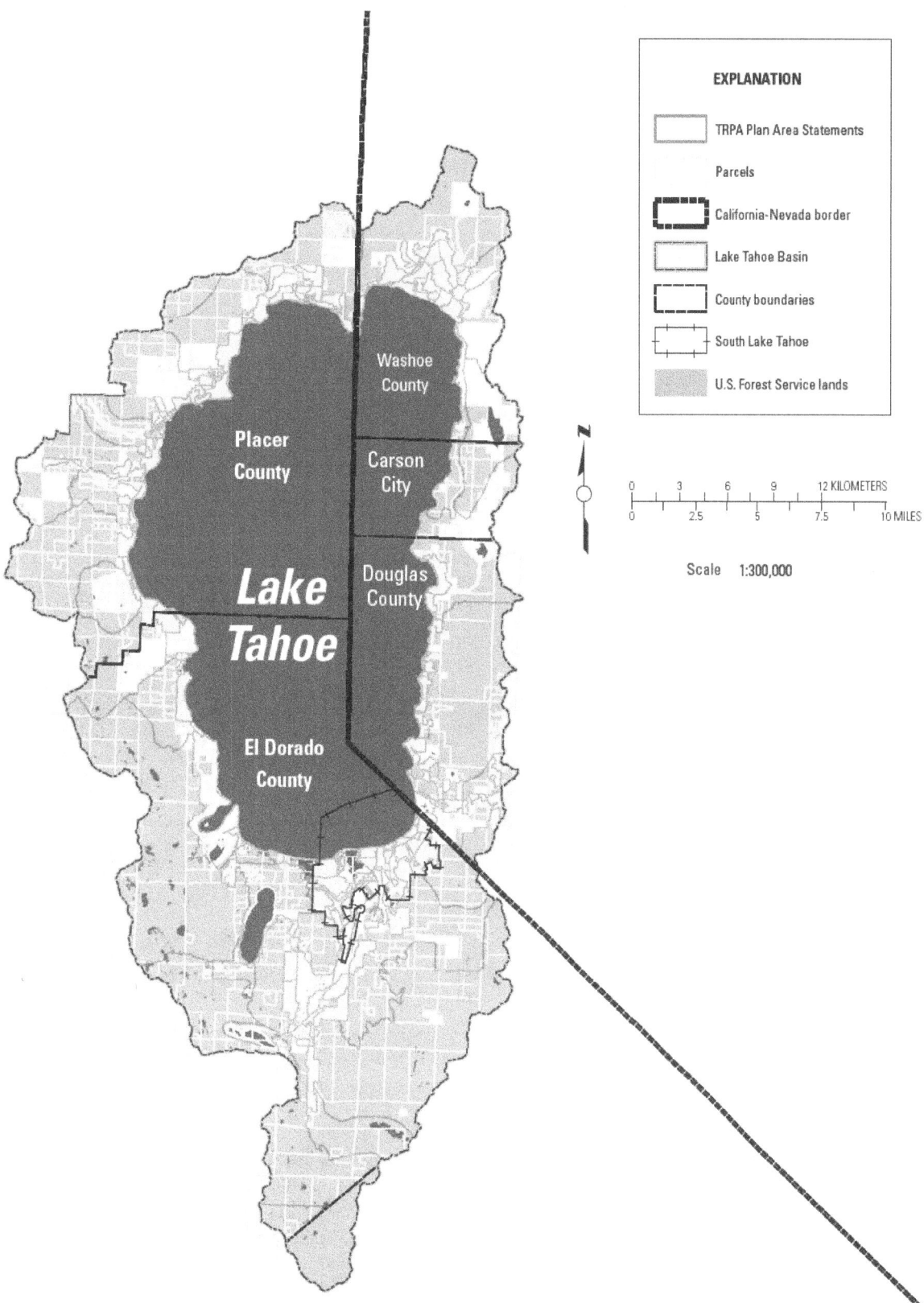

Figure 2. Map showing an overview of Lake Tahoe Basin and inputs to the Land Use Simulation Model (LUSM) such as parcels, Tahoe Regional Planning Agency (TRPA) plan area statements, the California-Nevada border, county and city boundaries, and U.S. Forest Service lands.

Table 6. Historical activities of land acquisition programs in the Lake Tahoe Basin—parcels acquired by agency per year, 1982 to 2007, and summary statistics by program.

[U.S. Forest Service (USFS) and California Tahoe Conservancy (CTC) data was obtained directly from the two agencies in 1999. Tahoe Regional Planning Agency (TRPA) data was obtained from TRPA in October 2008. Nevada Division of State Lands (NDSL) data was obtained directly from NDSL in March 2009]

Year	USFS	CTC	TRPA	NDSL
1982	8			
1983	216			
1984	236			
1985	322			
1986	611	602		
1987	488	360		
1988	382	195		
1989	374	263		149
1990	280	180		158
1991	90	103		121
1992	92	80		34
1993	53	51		11
1994	44	92		9
1995	27	49		3
1996	14	58		1
1997	16	39		0
1998	9	31		7
1999	8	10		3
2000			2	2
2001			15	0
2002			39	0
2003			47	0
2004			38	2
2005			41	
2006			20	
2007			6	
Total	3,270	2,113	208	500
Minimum	8	10	2	0
Maximum	611	602	47	158
Median per year	91.0	86.0	29.0	3.0
Standard deviation	189.5	163.9	17.4	56.3

the writing of this report, table 7 details some of the categorical differences in the proposed zones. In the future, the new zoning map could be used as an additional or substitute input to the LUSM.

Another aspect of TRPA's land management and regulations is defined by its Code of Ordinances (TRPA, 2008c). Many of its Chapters are relevant to the development of the LUSM and have been referenced during the definition of decision rules and variables, and the construction of the model's

logic and execution. Three geographic scales and two mapping systems are particularly important to the actions taken by TRPA within the urban growth boundary. The three geographic scales include (1) the previously mentioned regional plan, (2) the neighborhood-level Plan Area Statements (PASs which effectively include community plans), and (3) the parcel-level land-use zoning. The TRPA staff oversees and manages more than 270 PASs and community plans (Lief Larson, TRPA, oral commun., 2008). A particular PAS provides detailed plans for

Table 7. Comparison between the 1987 regional plan and the proposed transect zoning map of the Tahoe Regional Planning Agency.

1987 regional plan zones	Proposed transect zones
Commercial/public service	Neighborhood center
Tourist	Town center
Residential	Tourist center
Recreation	Low-density residential
Conservation	High-density residential
	Special districts
	Developed recreation
	General forest and parkland
	Roadless area
	Wilderness

a specific area in written form that include land-use policies and regulations for that PAS. All projects and activities are regulated by the provisions of the PAS, and redevelopment efforts must be consistent with them as well. Each one of the more than 60,000 parcels in the basin is zoned with one of the 149 categories of permissible land uses (see Forney and Oldham, 2011, for full list). As described previously, the two land-capability mapping systems that constrict the potential fate of a given parcel are (1) the BLCS and (2) the IPES.

Allocations and Commodities

In the context of the management systems and planning maps as described above, one of the primary land-planning tools available to TRPA is their allocations, or a permit to develop a particular property. When coupled with a development right, the allocation can become one of three types of "commodities"—residential unit of use, commercial floor area, and tourist accommodation units. In conjunction with the four acquisition programs for open space, the fourth commodity (in other words, fate of a given parcel) is "retired," which requires the purchase of the parcel's development right. Figure 3 outlines the components, linkages and general mechanisms of the allocations and their framework for the LUSM that will be discussed further in the Methods Section.

Although the zoning system has much greater specificity (Forney and Oldham, 2011), the residential allocation pools used by TRPA and the LUSM include single-family dwellings (SFD, code 1011) and multifamily dwellings (MFD, codes 1005, 1006 and 1007). Any residential allocations not used in a given year by the various jurisdictions end up in the residential allocation rollover pool, which makes them available for moderate income housing or sensitive-lot retirement. Table 8 provides historical records of the number of allocations, uses, and rollover by jurisdiction, and table 9 aggregates those historical records for the basin by year.

A nuance of the residential allocation pool related to Placer County is related to the Vacant Lot Equation (TRPA Code of Ordinances; TRPA, 2008c) and the IPES threshold. As opposed to the other counties, the IPES threshold for Placer County—currently at 726, below which building is not allowed—has not moved since the inception of the program in 1987. To account for—and eventually satisfy—this fixed threshold, the LUSM does not allow for development of parcels below the 726 threshold until 80 percent of the remaining sensitive lots are retired, thereby satisfying of the Vacant Lot Equation. The form of the Vacant Lot Equation is a ratio of the sensitive lots in rotation divided by the total number of lots in that county scored by the IPES field team in the late 1980's, which in the case of Placer County was 1,667. Consequently, to satisfy the equation, 20 percent of the original 1,667 lots need to remain in circulation in Placer County, thus 1,334 parcels need to be retired. To do so, the LUSM weights the selection of parcels towards sensitive-lot retirement (in other words, between 1 and 725 in IPES score) by the acquisition programs active in that county. Then, after the Vacant Lot Equation has been satisfied, the threshold drops to 1, and the LUSM releases the remaining parcels into the primary residential allocation pool of Placer County. In effect, there is a two-staged release of parcels in Placer County in the execution of the LUSM.

According to TRPA, approximately 74,322 square meters (m^2) of commercial floor area (all 3,000s codes, Forney and Oldham, 2011) currently exist in the basin. Under the current regional plan, approximately 18,580 m^2 of floor area are available for allocation and development. In the new Regional Plan Update, approximately 18,580 m^2 of additional floor area will be made available for a total of approximately 37,161 m^2. This additional amount, however, will be mostly provided to previously built-on parcels, and will be allocated to redevelop these lots. Targeting parcels for redevelopment in the model was considered, but a good method could not be derived (see Discussion Section for more). Primary commercial uses include retail, entertainment, services, light industrial, and wholesale/storage.

The primary tourist accommodation units (TAU) are bed-and-breakfast facilities, hotels, motels and other transient dwelling units, and time-share units of various design types. Besides providing lodging for visitors, tourist accommodations include secondary accessories such as garages, pools, tennis courts, bars and restaurants, maintenance facilities, gymnasiums, meeting rooms, child-care facilities, emergency facilities, and restricted gaming (TRPA Code of Ordinances; TRPA, 2008c). The LUSM, however, only chooses if the parcel is designated as a hotel, and excludes bed and breakfast facilities, and time-share units. For the purposes of the LUSM, only one representative commercial category is needed, and we assume that the details of the layout and particular amenities (for example, hotel versus time share, secondary accessories, and layout) are decided later by the developer, land-use planners, architects, and landscape architects. Although the reworking of the regional plan may alter the number of TAUs available, the TAUs currently available are 78.

Table 8. Historical residential allocations given by the Tahoe Regional Planning Agency that have been used and rolled over by county and city jurisdiction for the Lake Tahoe Basin.

[Please note that the median value for the rollover pool was used as a default for the model. Data provided by Tahoe Regional Planning Agency. -, represent null values]

Year	Douglas County			Washoe County			El Dorado County			Placer County			City of South Lake Tahoe		
	Allocations	Used	Rollover pool	Allocations	Used	Rollover pool	Allocations	Used	Rollover pool	Allocations	Used	Rollover pool	Allocations	Used	Rollover pool
1987	18	-	-	67	-	-	107	-	-	103	-	-	-	-	-
1988	23	-	-	67	-	-	106	-	-	103	-	-	-	-	-
1989	23	-	-	45	-	-	61	-	-	103	-	-	-	-	-
1990	18	-	-	45	-	-	175	-	-	63	-	-	-	-	-
1991	18	-	-	45	-	-	174	-	-	63	-	-	-	-	-
1992	23	20	3	59	53	6	130	129	1	88	83	5	-	-	-
1993	23	17	6	59	52	7	130	127	3	88	68	20	-	-	-
1994	23	15	8	59	35	24	130	120	10	88	62	26	-	-	-
1995	23	22	1	59	48	11	130	111	19	88	76	12	-	-	-
1996	23	23	0	59	59	0	130	130	0	88	88	0	-	-	-
1997	23	22	1	59	53	6	92	80	12	88	50	38	38	30	8
1998	23	17	6	59	38	21	92	81	11	88	49	39	38	35	3
1999	23	17	6	59	20	39	92	85	7	88	54	34	38	35	3
2000	23	23	0	59	32	27	92	88	4	88	51	37	38	30	8
2001	22	22	0	59	47	12	92	86	6	88	55	33	38	35	3
2002	22	12	10	59	24	35	92	84	8	88	60	28	38	35	3
2003	13	13	0	37	17	20	111	111	0	46	46	0	41	41	0
2004	14	14	0	40	18	22	90	90	0	46	46	0	35	35	0
2005	12	12	0	34	15	19	83	82	1	50	50	0	29	29	0
2006	13	-	-	31	-	-	83	-	-	50	-	-	35	-	-
Total	403	249	41	1060	511	249	2192	1404	82	1595	838	272	368	305	28
Median	23	17	1	59	36.5	19.5	99	89	5	88	54.5	23	38	35	3
Mean	20	18	3	53	37	18	110	100	6	80	60	19	37	34	3
Std. Dev.	4.1	4.2	3.5	10.9	15.6	11.4	29.5	19.8	5.7	19.2	13.8	16.0	3.2	3.7	3.1

Table 9. Total historical residential allocations given by the Tahoe Regional Planning Agency that have been used and rolled over for the entire Lake Tahoe Basin on an annual basis.

[-, represent null values]

Year	1987	1988	1989	1990	1991	1992	1993	1994	1995	1996	1997	1998	1999	2000	2001	2002	2003	2004	2005	2006	Total	Median
Allocations	295	299	232	301	300	300	300	300	300	300	300	300	300	300	299	299	248	225	208	212	5618	300
Used	-	-	232	-	-	285	264	232	257	300	235	220	211	224	245	215	228	203	188	-	3307	230
Rollover Pool	-	-	-	-	-	15	36	68	43	0	65	80	89	76	54	84	20	22	20	-	672	48

Open space is defined as "land with no land coverage and maintained in a natural condition or landscaped condition consistent with Best Management Practices, such as, deed restricted properties and designated open space areas" (TRPA Code of Ordinances, chap. 18; TRPA, 2008c). The transfer of development rights (TDR) is a typical practice used in real estate land management and conservation planning, where private landowners voluntarily sell—in perpetuity and under legal contract—the types of development to which they are entitled. For example, a development right can be related to the available air space above a diminutive historic building in a high-rise urban environment or it can be a unit area of building footprint allowed under land-use zoning and planning regulations. In return for the TDR, the original owner can receive a payment from another landowner for the right to develop elsewhere (as needed in the case of building multifamily dwellings in the Lake Tahoe Basin) or a payment from land-acquisition programs to retire the land to open space. Another type of practice is conservation easements, where the rights for natural-resource extraction, such as agricultural production, timber harvesting, and (or) mining, are transferred to a government agency or land trust, and the land is encumbered with the land-use restriction. In return, the original property owner retains rights to the land for other uses and may receive a reduction in their State and Federal tax burden because of the perceived decrease in the market value and utility of their land.

For the LUSM, some specific considerations of TDRs and conservation easements are relevant to address. TDRs are primarily related to the ground-floor footprint of area consumed by a building on the landscape. Although TRPA does have a sensitive-lot retirement program subject to the allocations in the residential rollover pool (fig. 3), it is unable to hold lands in title. Instead, TRPA works in partnership with other acquisition programs in and around the basin. As discussed before, the other three acquisition programs are USFS (as funded by the Santini-Burton Act, Public Law 96-586, 1980), the CTC, and the NDSL. The first is a Federal program, and the latter two are State level programs. There are no land trusts. Consequently, the LUSM must address both the geographic domains of the programs and account for the particular open-space retirements of specific acquisition programs.

Redevelopment, Mixed Use and Compact, Form-Based Design

A recent TRPA initiative focuses on targeted areas of infill, mixed-use, and compact redevelopment in the limited urban areas of the basin (TRPA, 2008b). This includes a focus on form-based zoning and design, which includes prescriptive and possible discretionary rules applied to particular development sites (in other words, a single-family residence area combined with offices and retail areas as long as the latter conformed to such code requirements as setbacks, lot coverage, and height). The research of the past 20 years has shown that 70 percent of the source inputs reducing lake clarity derive from fine sediment runoff of urban centers, which are only 1.4

percent of the regional land area (TRPA, 2008b). To facilitate the planning focus on these areas, a pilot program called the Lake Tahoe Community Enhancement Program (CEP) was established in the summer of 2007. The "focus of the CEP is to implement projects that demonstrate substantial environmental, as well as, social and economic benefits through mixed-use development projects on existing disturbed and / or underutilized sites" (TRPA, 2008c). The place-based program goals and objectives include create/enhance mixed-use community centers, create multimodal transit for the future, strengthen and create gathering places and economic centers, promote projects that construct threshold-related environmental improvements, promote transfer of development rights for environmental benefit, rehabilitate substandard development[8], and permit and review process improvements and coordination (TRPA, 2008b). Although many of the CEP project selection criteria are at a scale, resolution and grain that are beyond the purview of the LUSM, some of them are relevant to consider. They include additional allocations of commercial floor area, tourist accommodation units and multiresidential bonus units, density-focused priorities, consistency with proposed land-use zoning and codes, compatibility with neighborhood characteristics, forest management, and reduction of fuel loads in the WUI.

Other Agency Jurisdictions

As mentioned previously, three other agencies have jurisdiction and management directives in the Lake Tahoe Basin, namely the USFS, LRWQCB, and NDEP. The USFS has a subregion specifically designated to the Lake Tahoe Basin, called the Lake Tahoe Basin Management Unit (LTBMU). With the majority of their lands being outside the urban growth centers, the LTBMU manages approximately 80 percent of the lands in the Lake Tahoe Basin and represents the major landholder. Much of these lands are on the periphery of the basin, but some are integrated with the urban and mixed-use elements of the landscape in the WUI (fig. 2). For LRWQCB and NDEP, however, the Lake Tahoe Basin represents a small part of Region 6 and the State of Nevada, respectively. As part of their program goals and management directives, LRWQCB and NDEP collaborated on the development of the TMDL technical report (LRWQCB and NDEP, 2009) and subsidiary recommendations, actions, investigations and analyses.

Pathway 2007

Pathway 2007 was envisioned as a means for the four agencies to collaborate on updating resource-management plans and to create a unified plan. The plan is meant to guide "land management, resource management and environmental

---

[8]Further described as underutilized, disturbed, blighted, over-covered, and (or) brownfield sites.

regulations over the next 20 years" (Pathway, 2007). The plan is meant to address development type and pressure, lake clarity, forest health, water quality and recreation, reduction of the threat of catastrophic wildfires, and balancing economic development to support the quality of life with preserving Tahoe beauty for residents and visitors alike. Between November 2005 and April 2008, more than 25 meetings or "forums" were held to implement the collaborative effort (Roberts and Reuter, 2007).

Five Counties and One City

Below the agencies and land-acquisition programs in the hierarchy of land management in the basin are six jurisdictions[9], including five counties and one city (fig. 2). They are Placer County (California), El Dorado County (California), Washoe County (Nevada), Carson City County (Nevada), Douglas County (Nevada), and the City of South Lake Tahoe (California).[10] These jurisdictions are responsible for a wide variety of duties, including building and burning permits on specific parcels and managing multiple parks and recreation sites. These jurisdictions receive allocations from TRPA, partially manage the allocation and building permits and inspection processes, and decide on locations for the allocations through the execution of development rights. At the end of the year, if allocations have not been used by the jurisdictions, they are returned to TRPA and enter into the residential rollover pool (fig. 3). For the purposes of the LUSM, the pattern of allocation that would be decided on a case-by-case basis by the counties and city is distributed stochastically within a PAS. The distribution of allocations is subject to the constraints of the allowed and special uses, the densities of the neighborhoods, the size of the parcel, and the physical limitations of the BLCS and IPES systems and preservation of SEZs (fig. 3).

Materials, Data Sources, and Conversions

The majority of the LUSM development was in the ArcGIS environment (version 9.2) and Python (version 2.4). Once the data was preprocessed and compiled, the datasets were exported to the open source PostGreSQL format. The first stage was to prepare the data in Microsoft Excel and ArcGIS environments and execute the model in Python, and

the second stage was to serve the model through the Internet and Servoy© (a Web-based deployment platform that depends on Javascript and proprietary licensing). This process was meant to align with the data management practices of TRPA so they could use the LUSM (and serve it publicly through the Internet, if they wished), while still maintaining the ability to collect and adjust new information from various sources to refresh the database, such as regular parcel-map updates from various county assessor's offices. After final updates were completed, the model was debugged, the unexposed[11] decision rules that adhere to the TRPA Code of Ordinances and the default values were vetted, and the solution was migrated to Servoy©. Software platforms for the Internet tool are Python for Windows, PyGreSQL 3.8.1 database wrapper for Python 2.4, and PostGreSQL Version 8.1.11. The password-protected Internet tool is housed currently on Tahoe Integrated Information Management System servers (http://www.tiims.org/Science-Research/Environmental-Modeling/TDSS.aspx).

In terms of the data used, table 10 provides their name, source, year, scale, and other relevant details. In the table, the primary input datasets to the LUSM are in bold. Other datasets were used to add necessary fields for the development of the model's database, its operation, and to create figures for this report. The majority of the data began in the ArcGIS environment, and the essential integrator or "hook" to link the various datasets together is the Assessor's Parcel Number (APN). Each parcel has a singular and unique APN. Some preprocessing was required to get the necessary data in the right format. This included a spatial join of the Soils Layer—and its BLCS derivative score—to the Assessor's Parcel Map. Because of the particularities of the processing routine and the differing geometries of the Soils and Parcels layer, the three standard match options (intersects, contains, and is within) of ArcGIS were run independently, then merged into a comprehensive parcel dataset for the model. Preprocessing also included extracting and compiling various definitions of development density by land-use code from the written PASs and community plans (TRPA, 2008a). Furthermore, the jurisdictions of the parcel dataset did not include the City of South Lake Tahoe, so it was necessary to create an exclusive set of parcels only related to that geographic area. The "Place Name" boundary for the City of South Lake Tahoe of the U.S. Census Bureau, did not geographically align with the linework of the parcel dataset provided by TRPA, so manual interpretation and expert opinion were used to determine the exact parcels within the city limits.

---

[9]A seventh jurisdiction exists, Alpine County, at the southernmost end of the Lake Tahoe Basin, with the majority of the county being outside of the watershed. It is not included in the LUSM, as we were not given its parcel information, it is near the ridgeline of the watershed far from the lake, it does not readily contribute runoff or pollutant loads to the lake, and its land is effectively unbuildable.

[10]The City of South Lake Tahoe was incorporated in 1965.

---

[11]The unexposed decision rules are unavailable to the average user and require administrative privileges, as they are strict rules taken directly from TRPA's Code of Ordinances (TRPA, 2008c).

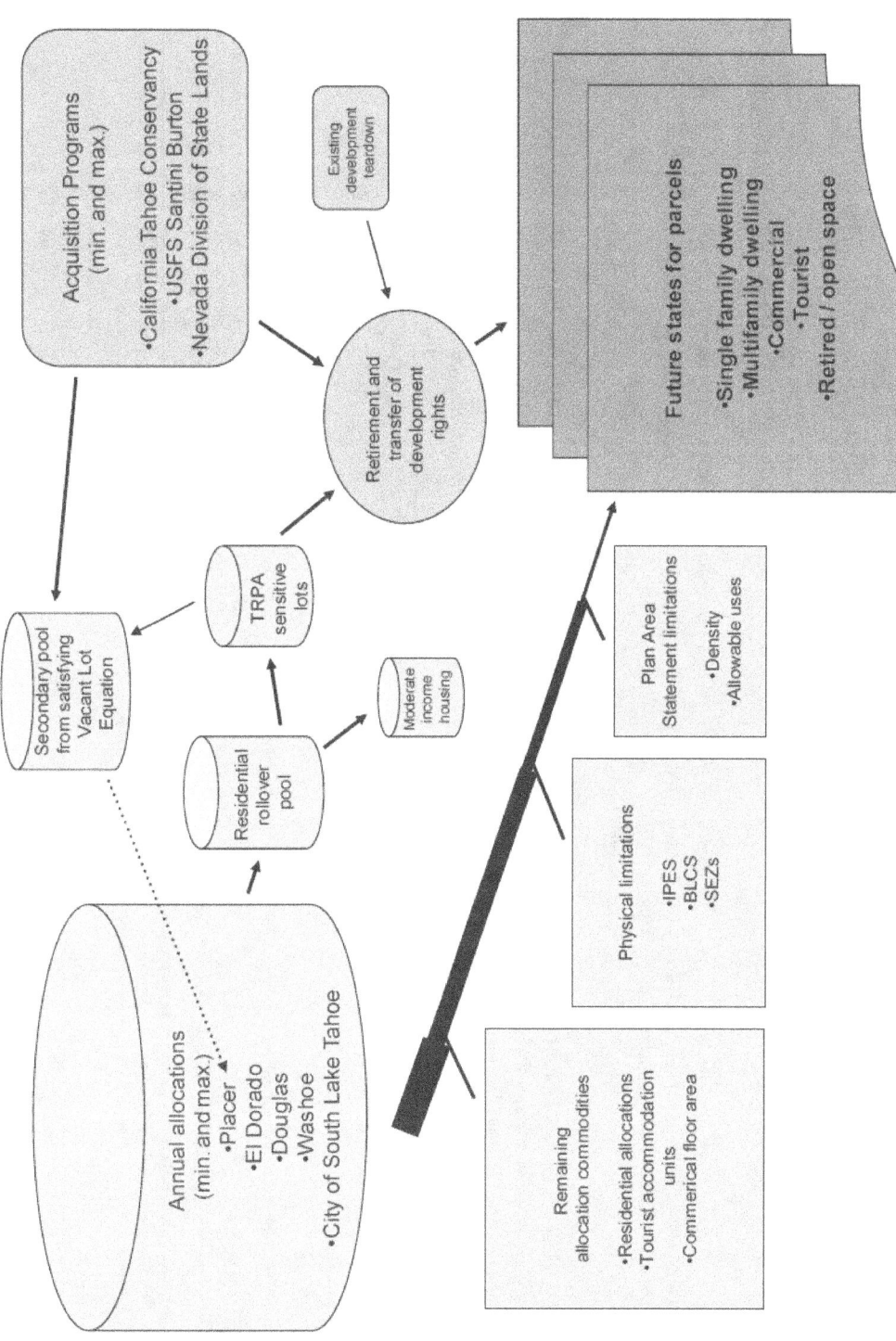

Figure 3. Conceptual schematic for the Land Use Simulation Model (LUSM). The blue components represent Tahoe Regional Planning Agency (TRPA) controlled land-allocation pools, the green components represent land-acquisition program allocation pools and the creation of development right transfers, and the yellow components represent the spatial, physical, and neighborhood limitations of the model execution that are placed on the potential future states of an individual parcel (in orange). The dashed line represents the fact that the Vacant Lot Equation (see text) applies only to Placer County. The Annual Allocations and Acquisition Programs (and their associated min and max values) relate to the formulation of the LUSM and the user's options for initial conditions. IPES, Individual Parcel Evaluation System; BLCS, Bailey Land Capability System; SEZ, Stream Environment Zone. min., minimum, max., maximum.

Table 10. Description of geospatial datasets used in the Land Use Simulation Model (LUSM) and this report.

[TRPA, Tahoe Regional Planning Agency; APN, assessor's parcel number; NRCS, Natural Resources Conservation Service; BLCS, Bailey Land Capability System; SEZ, Stream Environment Zone; USGS, U.S. Geological Survey; TEGIS, Tahoe Environmental Geographic Information System]

Dataset name	Source	Published[1]	Scale	Notes or comments
Assessor's Parcel Map[2]	TRPA	July 2008	Insufficient metadata; 2006 version had +/- 3 m accuracy	Includes parcel APN, IPES scores, land-use zoning, ownership
Plan Area Statements[3]	TRPA	July 2008	Insufficient metadata	Boundary polygons for zoning districts.
Plan Area Statement Uses[4]	TRPA	July 2008	Insufficient metadata	Includes allowed and special uses. Density metrics added.
Soils layer	TRPA (revised NRCS)	NRCS Soil survey, 2007	1:24,000 (typically done by NRCS).	Includes BLCS scores, SEZ
General Ownership	TRPA	September, 2006	Insufficient metadata	Basic land ownership
State Boundaries	National Atlas of the United States (USGS)	June, 2006	1:2,000,000	Partitioned data into exclusive spatial sets
County Borders	TRPA	September 2006	Insufficient metadata	Partitioned data into exclusive spatial sets
City of South Lake Tahoe Border	2000 Place Names—Census Bureau[5]	March 2003	Insufficient metadata	Partitioned data into exclusive spatial sets
Hydrologic Boundary of Lake Tahoe Basin	TEGIS	September 2003	Unknown	Graphical display
Existing Vegetation Types	Center for Spatial Technologies and Remote Sensing, University of California, Davis	Dobrowski and others (2006)	Maximum resolution = 4 meters; metadata has vegetation class accuracy assessment	Graphical display, landscape characterization and metrics
Stream Network[6]	TEGIS	September 2003	Unknown	Graphical display
Watershed Boundaries	TEGIS	October 2003	Unknown	Graphical display

[1]Some datasets did not have the metadata available to determine the date they were published, so in some cases the date created, the last date modified, or the date provided were used.

[2]Bold datasets are used in the execution of the LUSM.

[3]Includes community plans.

[4]Includes community plans.

[5]The Census Bureau provides another type of dataset for cities called Urbanized Areas, which consist of densely settled territories with (1) core census-block groups that have a population density of at least 1,000 people/square mile and (2) surrounding census blocks have overall density of at least 500 people/square mile. For South Lake Tahoe, this boundary would have included areas like Meyers, the neighborhood off North Upper Truckee Road, and the neighborhood off South Upper Truckee Road. For the LUSM, however, we assumed that the place name boundary represented the extent of a city's jurisdiction.

[6]The older stream network from TEGIS was compared to the more current National Hydrography Dataset (NHD, downloaded Dec. 2008), and the older stream network consistently matched the spatial locations of the flow lines of the NHD. Furthermore, the older stream network had better line-work connectivity and included lower order streams and headwater creeks. As a data model, however, the NHD has greater functionality in terms of attribute comprehensiveness, flow routing, and linkage between surface-water features.

Table 11. Characterization of the parcel dataset for the Lake Tahoe Basin by geographic unit; categories include, jurisdiction, number of parcels, vacant parcels, and summary statistics of their areas.

[ha, hectares]

Geographic unit	No. of parcels	Vacant parcels	Min. area, ha	Max. area, ha	Total area, ha	Mean area, ha	Standard deviation, ha
Lake Tahoe Basin	60,376	6,326	0	1,692	86,836	1.4	23.7
Placer County	15,802	1,823	0	269	18,400	1.2	11.9
El Dorado County	14,819	2,161	0	1,516	42,885	2.9	34.8
Washoe County	8,892	283	0	1,523	7,778	0.9	23.4
Carson City	42	2	0.5	368	2,536	60.4	72.5
Douglas County	5,576	276	0	1,483	11,161	2.00	32.0
City of South Lake Tahoe	15,243	1,781	0	73	2,369	0.1	1.2

Data Assessment

One of the primary inputs to the LUSM is the Assessor's Parcel database. Table 11 summarizes the parcel dataset and its geographic distribution by jurisdiction. Although they overlap geographically, the summary statistics for El Dorado County and the City of South Lake Tahoe are exclusive sets. Another primary input to the LUSM is the PAS and associated PAS Uses attribute table. The PAS shapefile provided by TRPA specified 309 separate geographic units (including community plans), with a minimum size of 3,245 m², maximum size of more than 104,307,000 m², mean of approximately 2,639,000 m², and standard deviation of over 9,121,000 m². The PAS Uses data table detailed the "allowed" and "special use" designations for each land-use zoning code in a given PAS.

In conducting quality assurance and quality control on the various datasets provided to us, we reviewed a number of factors that resulted in several notable observations. From the parcel layer (Assessor's Parcel Map, table 10), Forney and Oldham (2011) and figure 2 provide a summary of the dataset. Certain attributes are highlighted in figure 2, as they are important to development of the LUSM (in other words, assessor's parcels, PASs from TRPA, city and county jurisdictions, and the current ownership of the USFS lands). The land-use codes used by TRPA have three hierarchical levels of specificity (for more details on levels, see Forney and Oldham, 2011). After discussions with TRPA and independent research and analysis of the three levels, it was determined that the LUSM would mostly focus on Level I commodities—the exceptions being single-family dwellings and multifamily dwellings. Two hundred ten parcels did not have a land-use code and could not be used in the LUSM. Of the 6,326 vacant parcels that conceivably could be included in the LUSM set, two parcels were excluded as a result of being in Carson City County, and 43 did not have a PAS identity that could be

linked to the PAS dataset. This left 6,281 vacant parcels entering the model. One of the biggest observed problems was that the IPES scores of all vacant parcels in Placer County have values of zero. This is curious, as it is known that 1,667 residential lots were scored at the inception of the program. Furthermore, in all counties, it is unclear if a value of zero for an IPES score for a parcel record indicates that (1) it actually has a score of zero, and is effectively unbuildable, or (2) that the parcel was never scored, and is effectively a null value.

In terms of the PAS boundaries (table 10, Plan Area Statements), problems were found such as PASs split by county boundaries, instances of unclear data such as discerning a null value from a default value, and discrepancies between the database files provided directly by TRPA and their on-line PAS reports. Although it has some discrepancies as well, the PAS allowed and special uses data (table 10, Plan Area Statement uses) were more useful as there was one record per PAS. Metrics of density, although specified in individual PAS reports—and of interest to TRPA staff as it reflects the evolution of how planning is conducted in the basin—were not included in the datasets provided by TRPA. As such, we culled the on-line documentation for the specific densities allowed in the planning documentation (TRPA, 2008c). Governed on a per unit area basis, the commodities included MFD, commercial uses, and TAUs. The densities are as follows:

- MFDs—depending on the PAS in which they are located—are as many as 3, 8, 12, 15, 20, and 40 units per acre.

- Commercial-use square footage is 25 percent of a parcel's area.

- TAUs are 15, 20, and 40 units per acre.

Methods, Model Design, and Logic

This section discusses topics relevant to the methods, model design, and logic of the LUSM. It includes discussion of design criteria for the decision support system, collaborator communication and feedback, modeling theory and approach, decision rules imbedded in the model, the development intention and retirement intention for a particular parcel, the context and translation of a generalized model framework to this particular purpose and logic, functional considerations, and assumptions of the LUSM.

Design Criteria for the Decision Support System

By definition, a decision support system (DSS) is an interactive, computer-based tool—or collection of tools—intended to enable decisionmakers to better use the information in data and models to identify and solve problems. The objective of using a DSS is both to improve (1) the process of decisionmaking and (2) the effectiveness from the outcomes of decisions. They are designed to assist in making decisions and support—rather than replace—managerial judgment (D'Erchia and others, 2002). An ideal, integrated suite of tools should have the following characteristics: interdisciplinary focus, dynamic temporal aspects, three-dimensional visualization, multiscaled in space, policy-relevant to the regional and local context, iterative and modifiable in application and design, additive with other tools, accessible to a wide range of users with clear documentation and features, and affordable to acquire and (or) use (Condon and others, 2009; National Science Foundation, 2009).

Requests to revise earlier TDSS models were proposed by TRPA to make the results more relevant to their needs in decisionmaking. They included (1) updating the model with new parcel data, (2) disaggregating it to an individual parcel level[12] and maintaining the APN as the key data identifier, (3) removing the hard coding of the development intention (to be discussed later) that severely limited the possibilities of future states for any given parcel, (4) adding the model's ability to include stochastic simulations and measures of uncertainty of development intention, (5) adding the uses designated by the PASs and their associated building density by zoning type, (6) removing the interpolated missing values of IPES for parcels and replacing them with BLCS scores in situations where relevant (in other words, never scored residential lots, commercial lots, and tourist lots), (7) outputting results to a site-specific physical location, in other words, putting them on a map, which enables linking them to geospatial analyses, other models and databases, (8) considering incorporating redevelopment into the model for the Regional Plan Update,

(9) producing probabilistic results of the likelihood of a parcel transitioning from its current vacant state to an alternate state, and (10) migrating the new database, model platform, and solution to another TRPA server for its long-term usage.

Collaborator Communication and Feedback

Because of the need to make simplifying assumptions, models of policy effects inherently require "consultation and participation" from early development through implementation (Yearley, 1999). As such, in addition to two team meetings and multiple emails, the development team conducted regular, biweekly conference calls from June 2008 to November 2008. Discussions and communication often revolved around several questions and topics, listed below. The answers to these questions were integral to the development of the model.

Determining the "What" and Amount of Allocation Pool(s)

- What are the numbers of permits for the basin? What are the numbers of permits per jurisdiction? What are the numbers of parcel retirements for the basin? What are the numbers of parcel retirements per jurisdiction?

- How many permits can be allocated in a given year? How do they vary by land-use class (especially, residential, commercial (which is allocated on a floor area basis and may be related to a community enhancement program), and tourist accommodation units?

- In each land-use class, how many permits remain to be distributed until the basin is built out?

- How does the allocation rollover pool work? Typically, how many unused allocations roll over from one year to the next?

- How far down the land-use code levels of hierarchy does TRPA want to simulate? What level of specificity of granularity does TRPA want to be able to analyze?

- How many parcels are retired in a given year? Do the retirement agencies (for example, CTC) have particular numerical targets for acquisitions?

- How does TRPA locate affordable housing parcels in their regional plans? Is this type of land-use something TRPA will want to include in the LUSM? Or are these types of decisions delegated to County jurisdictions?

- How does TRPA locate employee housing parcels in their regional plans? Is this type of land-use something TRPA will want to include in the LUSM? Or are these types of decisions delegated to county jurisdictions?

- What makes a parcel a good candidate to be used as a "development rights transfer" parcel?

---

[12]For the purposes of the development of the TMDL with LRWQCB, a previous version aggregated the output to the 184 subwatersheds of the Tahoe Basin.

- What parcel allocations are aggregated? Are parcel allocations aggregated between jurisdictions? Only within a jurisdiction? Do the allocations become one big allocation pool for the entire basin or do they remain segregated by some geographic unit (in other words, watersheds, counties, and other units)?

- What is the mechanism for selecting commercial and tourist accommodation units for redevelopment? Will TRPA use the selection metric that was tested with Douglas County assessor's data (Forney and others, 2008)? What is the pace of redevelopment expected to be?

Determining the "Where" to Develop and Retire

- How is the development or retirement intention of a particular parcel determined? Does the PAS determine its fate? Do community plans determine its fate?

- How close are parcels to the various focal points (for example, transportation centers, commercial districts, employment opportunities, Lake Tahoe, open space, recreation, fire-prone forests, biodiversity hotspots)? How does proximity to various focal points govern the fate of a parcel?

- Are certain land-cover types more likely to be developed or retired? Does a land-cover type occur in a clustered or diffuse pattern?

- Do the retirement agencies have spatially explicit targets of lots for acquisitions?

- Where are the sensitive lots that should be retired? Is that determined by IPES or BLCS?

- Where are the nonsensitive lots that can be built on? Is that determined by IPES or BLCS?

- How are development rights transferred? Are they transferred within jurisdictions? Between jurisdictions? Within watersheds? Randomly?

- Does TRPA have spatial priorities for development or redevelopment locations?

- What slopes can support building? Where are slopes so steep that they need to be retired?

- How is the neighborhood of a potential parcel characterized? For a given parcel, does its proximal and distal neighbors influence its fate?

- Where are certain land-use types within a PAS? What are their percentages of distribution across PASs?

- Should big parcels be preferentially retired? Should small, sliver parcels be preferentially retired?

- Are the developable parcels (residential) outside of stream environment zones? Has building ever occurred in stream environment zones?

- Are the already existing, retired vacant parcels within stream environment zones?

Additional Considerations

- Should an LUSM output be an attribute table with multiple future-state probabilities for each parcel? Or just a single, future state?

- Should the amount of allowable impervious/semipervious coverage be incorporated in the analysis? If so, how would this type metric conform to the reality of actual coverage on the ground?

- Should an additional allocation type be specifically included for those parcels in need of restoration or hard surface removal?

- Is it possible to produce the model and its outputs so that it meets the needs of TRPA and LRWQCB?

Modeling Theory and Approach

As questions and research problems become increasingly complex and interdependent, modelers and the models they produce need to address interactions between human influences, ecological processes, and landscape dynamics. Similar to how the tool called EvoLand (Guzy and others, 2008) is formulated, given a simplified, binary consideration of land uses between retirement and development, the choice exists between social and policy goals that favor retirement for the sake of environmental quality and health versus those that favor land-use development for the sake of its associated economic benefits. As is the case in the Lake Tahoe Basin, these choices and land-use changes impact diverse aspects of planning and managing the complexities of human and natural systems (Verburg and others, 2002; Bolte and others, 2006). To help characterize these complexities, an agent-based modeling approach (simulating actions of autonomous entities with an orientation on system-wide assessments) was selected for the LUSM, as parcels can be considered the actors of an agent-based model to represent human development choices (individual, institutional, or organizational structures) in the basin, which drive landscape change over time (Bolte and others, 2006). As the LUSM is an agent-based model, it differs from cellular automata models in a variety of ways, such as neighborhood functions, dispersion rates, and transition rules (Balzter and others, 1998; Candau, 2000; Dieztel and Clark, 2006; Engelen and others, 1995; Wu, 1998).

Models of urban growth and landscape change such as the LUSM play an important role in planning, visioning, and scenario generation (Dieztel and Clarke, 2006). Urban (2000) used dual modeling frameworks where tabular, parametric space for modeling was coupled with spatial, geographic space for output and analysis. Furthermore, spatial decision support, which could not happen without the translation of the LUSM results to a GIS, is one of the techniques that allow the incorporation of multiple-criteria decision models. With LUSM-based map output that can be compared to other map output, the user can consider decision options, rank priorities of criteria, and elicit map-derived heuristic knowledge (Jankowski and others, 2001). To achieve this—and to meet the TRPA's requests for improvements to the LUSM (as described above)—it was essential for the output to be at the parcel level and spatially represented on maps. Furthermore, many of the land-planning and management decisions in the basin occur at the parcel level of analysis (TRPA and LRWQCB, oral commun., February 2008).

Typically, simulations can do two of three things well (generality, precision, and realism), and the developer must prioritize and balance them (Swartzman and Kaluzny, 1987). Previous studies have shown the difficulty of incorporating the realism of neighborhood effects in land-use planning tools. Bernknopf and others (2003) constructed a linear programming framework that had difficulty capturing "many of the holistic, heuristic, and flexible considerations that guide real land management choices." Furthermore, although stated in terms of linear programming techniques, Stough and Wittington (1985) noted that the "inability to take account of interactions between land-use activities in different parcels . . . is particularly limiting for environmental problems." After discussions with TRPA and investigating their available data, we determined that a viable, realistic solution to incorporate neighborhood effects hinged on the PASs and community plans.[13] To assist with their planning efforts, TRPA maintains a list of more than 270 distinct records of planning units. Each unit governs the neighborhood's characteristics, as each unit has a written description that includes such components as its overall goals, illustrative plans, and land-use regulations, as well as transportation, conservation, recreation, and public service objectives and policies (TRPA, 2008a). These PAS units define allowed and special uses and often include quantitative measures of maximum density and intensity of use by zoning type. The availability of this spatial dataset provided a method to derive development intention accurately and link it to a neighborhood level of consistency and control, yet incorporate aspects of stochasticity to account for uncertainty of future events, without making the LUSM so random that it would lose its meaningful representation of relevant policies and regulations for land-management decisions. Given the choice, the LUSM prioritizes precision of parcels and realism of PASs over generality.

Simulation models that incorporate real-world landscapes through maps and spatial representation have been recognized

as being crucial for the development of management strategies that account for regional land-use and global change drivers (Dunning and others, 1995; Bolte and others, 2006). For example, Verburg and others (2002) expressed the need to distinguish among the drivers that determine the quantity of change from those that determine the location of change (similar to the collaborator communication and feedback described above). In the LUSM, the quantity of change is dictated by the jurisdiction-specific development and retirement allocations of the initial conditions (table 12) and the density allowance of each PAS, whereas the location of change is dictated by the PAS use allowances, IPES, BLCS, the presence of SEZ, and parcel size.

Dieztel and Clarke (2006) emphasized that establishing a suitable level to which input datasets are disaggregated and hierarchically classified is an important consideration, and they suggested that going beyond the binary urban-nonurban thematic class input has repercussions for a model's output and validity. The distinction of multiple land-use classes allows for the consideration of competition between land uses, which, in traditional land-use change models that focused on deforestation, was not possible (Verburg and others, 2002). For the development of the LUSM, we went beyond the binary classification; however, we did not go to the finest classification level as found in Forney and Oldham (2011), as such a task would have become too onerous, the historical patterns of land use did not warrant it (for example, only one parcel is zoned "1010: residential care"; Forney and Oldham, 2011), and the utility of the model would have been diluted for TRPA and other users.

In terms of projecting landscapes into an uncertain future, modeling such situations becomes problematic when compared to some traditional modeling techniques. For example, the ability to calibrate and validate one's model with observed data is impossible, as the future condition does not exist (Bolte and others, 2006). As a result, the need to use historical data as a reference case and bound or constrain the realm of plausible transitions to build more likely scenarios for the future is important to incorporate (Bolte and others, 2006; Verburg and others, 2002). Thus, the LUSM's default values and transition rates depend heavily on what has occurred in the past, the use of the PAS provides a spatially nested degree of constrained realism in particular neighborhoods that is assumed to persist into the future, and particular policies and their decision rules and constraints for a given actor or parcel systematically direct the demand for its possible fates.

To consider the concept of corroborating the future landscape alternatives is more appropriate than attempting the impossible task of validating them as to a future condition that has not yet happened (Turner and others, 2001). With two datasets from two time periods (say, one historical and one current), Verburg and others (2002)—in their work with the CLUE-S model—suggested that the Kappa statistic, which is a statistical measure of inter-land-use-pattern agreement between the future projection from historical data and the actual present day configuration, could be useful for corroborating across the multiple-resolution landscapes with complex spatial patterns. Alternatively, and overcoming a limitation of

---

[13]For the purposes of the model, the PASs and community plans were considered homogenous and equivalent planning units, and the mention of PASs in this report is meant to incorporate community plans.

Table 12. User variables, their descriptions, and the default values for the Land Use Simulation Model.

[SFD, single-family dwelling; MFD, multifamily dwelling; TAU, tourist accommodation unit; CTC, California Tahoe Conservancy; NDSL, Nevada Division of State Lands; TRPA, Tahoe Regional Planning Agency; USFS, U.S. Forest Service; PAS, Plan Area Statement; sq. ft., square feet; ac , acres]

Parameter or commodity name	Description or notes	Default value
	Development	
Douglas county—minimum	Spatial allocation by county of SFD	9
Douglas county—maximum	Spatial allocation by county of SFD	21
Washoe county—minimum	Spatial allocation by county of SFD	13
Washoe county—maximum	Spatial allocation by county of SFD	49
El Dorado county—minimum	Spatial allocation by county of SFD	27
El Dorado county—maximum	Spatial allocation by county of SFD	111
Placer county—minimum	Spatial allocation by county of SFD	18
Placer county—maximum	Spatial allocation by county of SFD	66
South Lake Tahoe—minimum	Spatial allocation by county of SFD	11
South Lake Tahoe—maximum	Spatial allocation by county of SFD	47
Multifamily dwelling—minimum	Spatial allocation for basin of MFD	30
Multifamily dwelling—Maximum	Spatial allocation for basin of MFD	50
Tourist accommodation unit—minimum	Spatial allocation for basin of TAU	30
Tourist accommodation unit—maximum	Spatial allocation for basin of TAU	50
Commercial area—total sq. ft.	Spatial allocation for basin of Commercial for the entire model run	160,000
	Retirement	
CTC—minimum	Spatial allocation within California	86
CTC—maximum	Spatial allocation within California	250
CTC—Priority	Weighting of preferentially sensitive lands to retire	0.5
NDSL—minimum	Spatial allocation within Nevada	3
NDSL—maximum	Spatial allocation within Nevada	59
NDSL—Priority	Weighting of preferentially sensitive lands to retire	0.5
TRPA—minimum	Spatial allocation for Tahoe Basin	29
TRPA—maximum	Spatial allocation for Tahoe Basin	46
TRPA—Priority	Weighting of preferentially sensitive lands to retire	0.5
USFS—minimum	Spatial allocation for Tahoe Basin	91
USFS—maximum	Spatial allocation for Tahoe Basin	189
USFS—Priority	Weighting of preferentially sensitive lands to retire	0.5
	Other	
Allocation rollover rool	Initial amount in the pool	48
Minimum size	Minimum size threshold (ac.)	0.1
Special use allowance	Inclusion of special uses as designated by individual PASs	No
Starting year	Beginning year of the model run	2009
Ending year	Final year of the model run	2028

the CLUE-S model approach, which requires land-use change history, corroborating future landscapes can be done through expert opinion, incorporation of stakeholder's opinion, and incorporation of measures of uncertainty, all three of which have been done in the LUSM. Expert and stakeholder's opinions were incorporated from the knowledge of the model developers and discussions with TRPA, and measures of uncertainty were incorporated by the use of probabilistic output and other statistical measures of the model's performance. By no means should the alternative future or the landscape's trajectory to be considered "truth." The purpose of the LUSM is to be a decision support tool for policy considerations and choices of management actions and strategies (Bolte and others, 2006), and to compare future land-use scenarios with other geospatial models and data on topics such as biodiversity hotspots, existing native land covers, and landscape stability and processes (in other words, fire regimes, erosion, and sedimentation patterns) (Verburg and others, 2002).

Decision Rules

The decision rules are meant to add transparency to assumptions and drivers of the allocations by the LUSM to the various agencies, programs and jurisdictions in the Lake Tahoe Basin. Some of the decision rules were defined in previous versions of the model (Hessenflow and Halsing, 2006). Through our discussions with TRPA staff and analysis of their Code of Ordinances, the decision rules and associated assumptions were refined and expanded such that:

- Allocations are distributed annually from TRPA to the counties and the City of South Lake Tahoe.

- At the end of a given year, unused residential allocations return to TRPA and enter the residential rollover pool.

- The residential rollover pool can be used in future years for moderate-income housing and (or) TRPA's Sensitive Lot Program

- PASs dictate the possible commercial allocations in a neighborhood and remain in force for the duration of the model's run.

- To incorporate the future's uncertainty, a parcel may receive multiple types of allocations with explicit probabilities.

- Each jurisdiction and retirement program can be seeded with an initial condition of minimum and maximum allocation targets for SFD for a given year.

- MFD, TAU, and commercial commodity targets are set for the entire basin.

- For a SFD, one allocation is required for a building permit.

- For a MFD, the number of allocations required must equal the number of units on the parcel (in other

words, a duplex needs two allocations, a fourplex needs four allocations).

- MFDs are the only commodity type that receives the development right transfer.[14]

- Assuming availability of adequate lands and capability, commercial parcels may have as much as 25 percent of their area covered.

- An IPES score of "0"and a BLCS score of "1b or 1c" makes a parcel unbuildable.

- An IPES score between "1 to 725" and a BLCS score of "1a, 2, or 3" deems a parcel to be sensitive land.

- An IPES score higher than 725 and a BLCS score higher than 3 deem a parcel to be less sensitive and more buildable land.

- Retirement agencies are interested in varying levels of land sensitivity and parcel sizes.

- A purchase of the four acquisition programs both puts the lot permanently into open space (also called retirement) and makes possible the transfer of its development right.

- Starting with the available vacant parcels in the basin, SFD, MFD, TAU, commercial and retired commodities are the only commodities included in the output of LUSM.

As outlined in conceptual schematic for the LUSM (fig. 3), a number of interdependencies between these decision rules, model components, functions, and routines were incorporated into the mechanics and coding of the LUSM (Forney and Oldham, 2011).

As discussed in the Lake Tahoe Land Management Section, a nuance of the mechanism of the residential parcel allocations in Placer County is associated with the Vacant Lot Equation. The Vacant Lot Equation requires that a certain number of parcels be retired before additional, more sensitive lots (as defined by IPES) are released to the public and made available for land-use change. In effect, there is a two-stage process in Placer County for the release of lots. During the first stage, only vacant parcels above the IPES threshold of 726 can be developed. At the same time, retirement activities are removing more sensitive parcels below the threshold. Once the ratio of the Vacant Lot Equation reaches the required 20 percent (TRPA Code of Ordinances), then the second stage is achieved as the threshold is removed and any parcel that has a score of 1 or higher is put into circulation as a potential parcel for change of state (in other words, development or retirement).

---

[14]Currently, TRPA does not track the linkage between the origin and destination of development right transfers. In the future, however, they may both maintain a record of the transferring parcels and regulate their transfer so that the development right is kept within certain jurisdictions or watersheds.

Table 13. Residential allocation performance table (Tahoe Regional Planning Agency, 2008c) used in the Land Use Simulation Model.

[n.a., not applicable; Nev., Nevada; Calif., California; CSLT, City of South Lake Tahoe; min., minimum; max , maximum]

Jurisdiction or county/city	Min. allocation with deductions	Deduction increments	Base allocation	Enhancement increments	Max. allowable enhancement
Douglas County, Nev.	9	−1	13	1	21
Washoe County, Nev.	13	−3	25	3	49
El Dorado County, Calif.	27	−7	55	7	111
CSLT, Calif.	11	−3	23	3	47
Placer County, Calif.	18	−4	34	4	66
Total	78	n.a.	150	n.a.	294

To consider the residential allocations further, the range of allocations each jurisdiction receives in a given year is characterized by table 13. The deduction and enhancement increments are defined by the jurisdictions' "bad" or "good" actions taken over the previous year. For example, in Douglas County, the base allocation is 13, the minimum allocation is 9, and the maximum allocation is 21. The five actions that shift the allocation up or down are as follows:

1. Given a sample of auditing, the monitoring of permits and compliance from inspections. The basic expectation is a passing score of 70 percent.

2. Receipt of water quality and air quality EIP project list and inclusion of their schedule of completion.

3. Receipt and approval by TRPA of an updated EIP component list and completion of projects.

4. Implementation of BMP retrofits and development of TRPA approved programs to assist with that implementation. Enhancement increments are increased by demonstrating adequate resource commitment to the program. Yearly targets need to be specified, and the baseline for implementation is 2004.[15]

5. Establishment of a 2003 baseline for Transit Level of Service (TLOS), which is monitored by the jurisdictions' transit operating funds to improve that level of service. Level of improvement is measured by criteria in the TLOS Guidelines Handbook.[16]

6. Assessment of these actions is based on seven criteria listed in chapter 33 of the TRPA Code of Ordinances.

Given input from the user, the LUSM accounts for these deduction and enhancement increments by providing a range of initial conditions (in other words, a minimum and maximum value). Although default values (table 12) are provided in the model from analysis of historic records, discussions with TRPA, and expert opinion, the user must make their determination of the input values that each jurisdiction ought to receive. To extend that logic, the user helps establish priorities for the importance that each allocation will have and the number of times it will be selected through the execution of the model, the initial target allocation levels of the various commodities (in other words, minimum and maximum numbers of TAU versus SFD versus commercial and other classifications), establish the preference at which the model selects parcels to transition, and the annual targets it tries to achieve.

Development Intention

The development intention is the fate of a parcel after it has been selected to transition from its current state (in other words, vacant) to a new state. This transition can occur in any year of the predefined timeframe of the simulation runs—the default run is 20 years to coincide with the Pathway 2007 effort. A parcel can be selected, and then transitioned, only once. A previous version of the model (Duffie and others, 2004) used historic records of temporal change from 1998 to 2003 to define and hardcode the development intention. Although this method is useful and has theoretical basis in determining urbanization rates (for example, Dietzel and Clarke, 2006, used 2- and 4-year time steps), the accuracy and comprehensiveness of these earlier parcel datasets is suspect, thereby bringing into question the validity of their derived values. Furthermore, this method must assume that the sample set provides all relevant market transactions and allocations within that time step and that these are representative enough to extrapolate to a longer time horizon.

Different techniques for determining the parcel's development intention were considered, including its land-use code and an assessment of the distribution of existing land-use codes. The existing land-use code would not suffice, as it

---

[15]This combines two of the criteria (33.2.B.5.d and 33.2.B.5.e) from the TRPA Code of Ordinances.

[16]This combines two of the criteria (33.2.B.5.f and 33.2.B.5.g) from the TRPA Code of Ordinances.

would not inform the future state, for example, a vacant parcel has only one code for the present state, namely vacant. The distribution of existing land-use codes showed some promise as it reflects the landscape as it has evolved over time to its current state; therefore, it is a past predictor and rationally justified. This technique had three identified limitations—(1) as in financial analysis and investment theory (Rosenberg and Guy, 1976), historical values are not completely accurate as future forecasters in that they do not incorporate an adequate measure of uncertainty; (2) the definition and value of the asset and its uncertainty (in other words, average land use of a given type of parcel or average rate of transition over time from vacant to a particular land-use type) would be unclear (Rosenberg and Guy, 1976); and (3) the context of the parcel, namely its surrounding neighborhood and physical landscape would not be incorporated into a parcel's fate.

Retirement Intention

The decision rules determining the intention for the retirement of parcels of land are fairly well defined by the TRPA Code of Ordinances. In addition, the "gray" literature and mission goals of the basin's acquisition programs (CTC, 2005; NDSL, 2008; and USFS, 2008) helped to refine the decision rules, and develop additional variables to incorporate the preferences of a wider stakeholder group of potential model users. In terms of the TRPA Code of Ordinances, the primary determinant is land capability and the BLCS and IPES. Chapter 34 of TRPA Code of Ordinances indicates that according to the BLCS, sensitive lands—and therefore lands available for acquisition and retirement—are in classes 1a, 1b, 1c, 2, 3 (table 3) or SEZs, and according to IPES are below the IPES threshold (either 1 or 725, depending on the parcel being located outside or inside Placer County, respectively). For the purposes of the LUSM, the authors attempted to cross-walk the BLCS with the IPES (table 3), which addresses those parcels never receiving an IPES score (in other words, residential parcels developed before 1986, and tourist- and commercially zoned parcels) and those that have inadequate IPES data. In terms of the basin's acquisition programs, expanded decision rules included size and sensitivity.

Logic and Functional Considerations

The LUSM is designed as a constrained, stochastic simulation model (Forney and Oldham, 2011). This was done on the basis of TRPA requests, data availability, consideration of land-use modeling theory, and the fact that deterministic outcomes would be misleading about a parcel's fate. Future scenarios are important to characterize with an adequate balance of realism and uncertainty. Considering the complexity of the construction of the model, a framework adapted from Turner and others (2001) was a useful organizational and communication tool for developers, stakeholders, and readers. Table 14 provides generalized terms and definitions of model

components and procedures, as well as the particular meaning to the relevant components of the LUSM.

Each commodity type has certain variable values and decision rules that govern its behavior, many of which can be set by the user (table 12). Most of the default values were derived from historic averages for parcel retirements, records of allocations and the rollover pool from TRPA, and the Residential Allocation Performance table (tables 6, 8, and 13, respectively), communication with TRPA staff, and the decisions of the model developers. The minimum retirement value was based on the median and the maximum was based on the median plus one standard deviation. Note that the minimum retirement value also includes spatial allocations by jurisdiction as well as commodity allocations by type. For all allocations by jurisdiction, commodity types, and retirement actions, a minimum and maximum number of parcels is allocated on an annual basis. The model selects an allocation value from a normally distributed, bell-shaped curve between the minimum and maximum values for a given year.

Some special considerations are required for the development commodities. First, commercial lots that are larger than the allocated value can "clog the pipeline" as the model executes and selects parcels, thereby blocking the ability to achieve the total allocation and reducing the amount of commercial development at the end of the model run. Therefore, code was developed to check if the commercial lot in question is simply too large to ever meet the designated commercial square footage value—if so, it was moved to the next year's list of parcels to allow for other parcels to be considered. This provided smaller commercial lots to receive an allocation instead. Second, because of the basin-wide consideration of MFD and TAU, we used a sequential pool approach that rolled over from year to year. For those two commodities, as their allocation targets execute on an annual basis, two procedures occur—(1) for any qualified parcel that bumps up against the allocation limit for units in a given year, it is first in line for selection in the following year, and (2) any remaining, unused allocation from a given year is added to the following year's allocation pool, thereby creating the possible situation where the maximum value for the following year can be temporarily greater than the original value input by the user.

In terms of parcel retirement, the model initially assumes that any vacant parcel could be retired. During the execution of the model, however, certain conditions need to be met. For any parcel without an IPES score, a BLCS score of 1b or 1c makes it undevelopable and therefore a suitable candidate for retirement.[17] For any SFD parcel with an IPES score of 0, it is undevelopable and therefore is a suitable candidate for

---

[17] For commercially zoned parcels, the TPRA Code of Ordinances is even more restrictive in that BLCS classes 1 (a, b, and c), 2, 3 or SEZ may not receive commercial floor area except in special circumstances related to transfers from more sensitive lands or if they are in community plans with SEZ restoration projects (TPRA Code of Ordinances, chap. 33.3.A.3). This level of detail, however, was not included in the development of the LUSM logic and functionality.

Table 14. Organizational framework for the Land Use Simulation Model (LUSM) (adapted from Turner and others, 2001).

[Framework includes general terms, their definitions, and how they apply to the LUSM. IPES, Individual Parcel Evaluation System; BLCS, Bailey Land Capability System; PAS, Plan Area Statement]

General term	Definition	Particular LUSM applications or examples
Parameter	Constant or coefficient that does not change in the model.	Jurisdiction, IPES score, BLCS score, parcel size.
Variable	Quantity that assumes different values in the model.	Min/max allocations, minimum size, min/max retirements.
State variable	Major elements of the model whose rate of change are given by differential equations.	Changes in allocation values/year.
Initial conditions	Values of the state variables at the beginning of the simulation.	Default values.
Forcing function, driving variable	Function or variable of an external nature that influences the state of the system but is not influenced by the system.	Land-use codes, PAS, vacant lot equation.
Output variables	Variables that are computed within the model and produced as results.	Transition probabilities, allocations/commodity/year.
Sink	Compartment in the model into which material or flow goes but from which it does not return.	Fate of parcels.
Source	Compartment from which the material flowing in the model flows but to which it does not return.	Number of vacant parcels in 2007.
Calibration	Process of changing model parameters to obtain an improved fit of the model output to empirical data.	No empirical data was available for comparison. Adopted a method to test model stability and increase in precision through increasing model iterations and analyzing coefficients of variation
Corroboration	Process of determining whether a model agrees with the available data about the system being studied.	Difficult. Checked to see that likely transitions "made sense" and were consistent with the allowed uses in a PAS.
Sensitivity analysis	Methods for examining the sensitivity of model behavior because of changes in variable.	Plan to conduct local sensitivity analysis by changing one variable at a time.
Verification	Process of checking the model code for consistency and accuracy in its representation of model equations or relations.	Four phases: producing, examining and reexamining output, developing model testing variables, extensive commenting in the code for explanation, migrating model and databases to new computer platforms and different users for replication of execution and functionality.

retirement. As described previously, Placer County has the special condition of the satisfaction of the Vacant Lot Equation, so until an adequate number of lots are retired in the County, the undevelopable threshold—and therefore a suitable candidate for retirement—is 725 or lower. This special circumstance is reflected in the model with less sensitive parcels are used up before any of the more sensitive parcels are released, thereby creating a secondary wave of available parcels with IPES scores between 1 and 725.

Although Chapter 34 of the TRPA Code of Ordinances indicates that parcels with BLCS scores of 1a, 2, or 3 are also deemed sensitive and available for retirement, we included a weighting scheme in the LUSM. This allowed the user to weight the selection of the four land-acquisition programs for priority and secondary land-sensitivity classes with the parameter *Priority*. This priority weighting, from 0 to 1, is like a sliding scale of how "hungry" the model is in its selection of sensitive lots as discussed in the Land Capability Section.

As for the other variables in table 12, the *Allocation Rollover Pool* relates to allocations not exercised or built on by local jurisdictions within a given year and that are returned to TRPA and placed in the pool. This rollover pool is explicitly available only for moderate-income housing or sensitive lot retirement in following years. To avoid the inclusion of "postage stamp" parcels and to allow for the prioritization of larger parcels, the *Minimum Size* parameter was added to the model. The *Special Use Allowance* can be toggled to include a less restrictive set of uses (in other words, those designated "special" in the plans) in

Table 15. Fixed variables and decision rules that persist in the background of the Land Use Simulation Model and can be changed by select users.

[IPES, Individual Parcel Evaluation System; MFD, multifamily dwelling; PAS, Plan Area Statement]

Parameter	Description or notes	Fixed value
Commercial area ratio	Ratio of parcel size to commercial footprint area.	25 percent
IPES threshold	Value at and below which parcels are unable to be developed for Placer County until the Vacant Lot Equation is satisfied.	725
Iterations	Number of iterations run by the model.	1,000
MFD units per acre	Density metric for MFD, varies by PAS.	18
Tourism units per acre	Density metric for tourist accommodations, varies by PAS.	30

all PASs. Although the data currently used in the model is from 2008, the *Starting Year* was included for future modifications to the model's database and to plan for the possibility of having multiple parcel datasets available (in other words, historical and (or) future) with which the database could be loaded and the model executed. Finally, the length of the model run can be determined by the *Ending Year*, with the default relating to the 20-year planning horizon of Pathway 2007.

In contrast to the variables that are available to the Internet user to change, fixed variables and decision rules exist and are only available to the administrators of the model such as TRPA and their authorized users (table 15). The *Commercial Area Ratio* and the *IPES threshold* come directly from the TRPA Code of Ordinances. The *Iterations* parameter is discussed further in Results Section, and the user can balance their need for precision in estimates against the computational demands of increased iterations. Both the *MFD Units Per Acre* and *Tourism Units Per Acre* were obtained from a review of the PAS descriptions and specifications, and they represent an upper bound of the values within those documents. Considering the discussions with TRPA and the priority in the TRPA Regional Plan Update towards increased density, the authors and model developers weighted higher default values for the density allowed by PAS (table 15). This was done for each commodity that is governed on a per unit area basis and has a variable value for densities by PAS, namely MFD and TAU. As stated previously, for MFD, the number of units that can be built is simply a function of the parcel's size, not the densities designated by the PAS. SFDs are limited to 1 per parcel, and commercial lots are limited to 25 percent of the area of the parcel. Although it can be changed by some users with administrative rights, the higher bound was chosen because TRPA expressed interest in investigating higher-density zoning plans in their update of the regional plan.

Assumptions

Considering the fact that the model is projecting into the future, it is impossible to calibrate and validate it in a traditional sense. Typical measures of model accuracy (in other words, explanatory power, statistical significance) are

irrelevant, as no future landscape exists that can determine if the outcome is accurate. Therefore, the model can only be corroborated (Turner and others, 2001). This is what makes it a scenario generation decision-support tool that can assist decisionmakers while considering the ramifications of present actions on future conditions.

In addition to the decision rules of the model (as presented earlier), the following represents the assumptions of the model that limits its scope and (or) that users need to be aware of as they conduct their own analyses with the tool:

- The needs and requirements of the TRPA take precedence over those of other past and potential collaborators.

- All data are provided by other institutions and agencies. This effort did not involve the collection of any new data (in other words, field-based study or laboratory experiments).

- Only parcels that are currently vacant are addressed in the model.

- Land-acquisition agencies can obtain and receive retired allocations in the future and do not have any capital or operational constraints that would keep them from purchasing the land at fair market value.

- Development allocations have interested parties who will implement the particular commodity (for example, TAU).

- Retired allocations include all legal and institutional practices that create open space (in other words, transfer of development rights, conservation easements, conversion to public lands, etc.).

- The parcel is the primary unit of analysis. Land-use and land-cover changes at finer resolution (in other words, subparcel) are not currently considered in the model.

Results

Users can access the DSS at the following Web site to run their own scenarios with the approval of TRPA: http://www.tiims.org/Science-Research/Environmental-Modeling/TDSS.aspx. Maintained by TRPA and TIIMS staff, the Web site includes a description of the background on the model. This section discusses the testing of the model, presentation of select results, the analysis of its output, and some limited observations about them.

Reasonability Analysis and Model Testing

Given the future projections of the results of the LUSM and the inability to traditionally calibrate and validate the model, the corroboration of output depends on two primary techniques—(1) visual assessments with ancillary datasets and (2) analysis of the coefficients of variation (CV). The visual assessments of output indicated reasonable outcomes associated with the default values of the model. Given their PAS and variation within it, size, and IPES and BLCS scores, vacant parcels transitioned to likely commodities. Around the basin, spot checks were made for each commodity, in each jurisdiction, and the model provided expected and intuitive results for parcels that are currently vacant.

In stochastic simulation, output values are expected to vary without making changes to the initial conditions of the parameter values. Furthermore, if the number of iterations increases without changing the input values, one would expect convergence towards certain values and a decrease in variation around these values. We used the CV to quantify the precision of the LUSM. Table 16 provides CVs by allocation type from 10 runs of the following iterations: 2, 10, 100, 500, 1,000, and 10,000. Figure 4 shows the same results in graphical form, with the addition of three power regression equations and their respective explanatory power.

Select Model Output: Tables and Maps

The model outputs seven comma-delimited ASCII tables. The seven tables serve four primary purposes—(1) two for producing probabilities of transition per land use for a given parcel, (2) one for producing the likelihood of an allocation occurring in a given jurisdiction or parcel retirement program, (3) three for assessing the rate at which parcels are consumed by land use, jurisdiction, and retirement programs, and (4) one for clustering the land uses into particular categories to assess their frequency of occurrence. For examples and further discussion of the seven table outputs, see Forney and Oldham (2011).

To track the rate of change in allocation pools over time, tables 17 and 18 represent the number of transitions of a certain land-use type that occur per year and the number of candidate vacant parcels left in a given jurisdiction for any given year, respectively. Note that the output of the model provides allocations of half parcels (in other words, 0.5), so those have been rounded down as half a parcel is not possible to allocate.

Using the default values for the model (table 12), and results from model simulation with 500 iterations, figures 5 through 11 present the spatially explicit, parcel-based probabilities of transition for the commodities of interest. Note in the figure explanations the number of parcels that end up in each class of transition probabilities. Although the lowest probability class includes low values outside of the following classification method, the three transition probability classes are defined by the Jenks natural break classification method (Jenks, 1967). The Jenks method seeks to minimize each class's average deviation from the class mean, while maximizing each class's deviation from the means of other groups. Table 19 presents a consolidated view of the number of parcels per commodity per transition probability class. Note that the probabilities transition classes have been simplified from numeric values to low, medium, and high categories for ease of comparison among commodities.

Figures 5 through 11 map the spatial distribution and concentration of the various commodities. As expected, for all commodities, the location of vacant parcel transitions occurs in relatively close proximity to Lake Tahoe. In figure 5, however, retired parcels are widely distributed around the basin and occur on locations higher up the hillslopes and even at the basin's rim. Furthermore, retired parcels tend to be larger in size. In figure 6, parcels tend to be closer to the lake, in PASs that allow residential and commercial allocations, and on smaller parcels. In figures 7 through 9, the parcels are clustered in certain areas such as the City of South Lake Tahoe, Kings Beach, Incline Village, and Tahoma. In those locations, the number of units of MFDs increased as the frequency of occurrence decreased. In figure 10, the few TAU parcels are concentrated in the primary tourist towns of the basin, namely City of South Lake Tahoe, Carnelian Bay and Kings Beach, Incline Village, Tahoe City, and Tahoma. In figure 11, the locations of the few commercial allocations mimic the spatial distribution of the TAU parcels but also include locations between Carnelian Bay and Tahoe City and the Highway 50 corridor headed south from the City of South Lake Tahoe.

The results of the LUSM can be compared to other spatial datasets to determine potential impacts to such natural resource management concerns as land cover, wildlife biodiversity, risk of and vulnerability to fire, and locations of erosion potential. As an example comparing geospatial model outputs and datasets, a first order analysis using the spatial data of figure 1 and figures 5 through 11 can be conducted to create a matrix of transitions that indicates the land cover that would change to a particular land use for a given scenario (in this case, the scenario of the model's default values). The results of this analysis are the area of certain land-cover classes that transitioned to a different land use (table 20). These results required the simplification of the transition probability class results of figures 5 through 11 to use only the highest class.

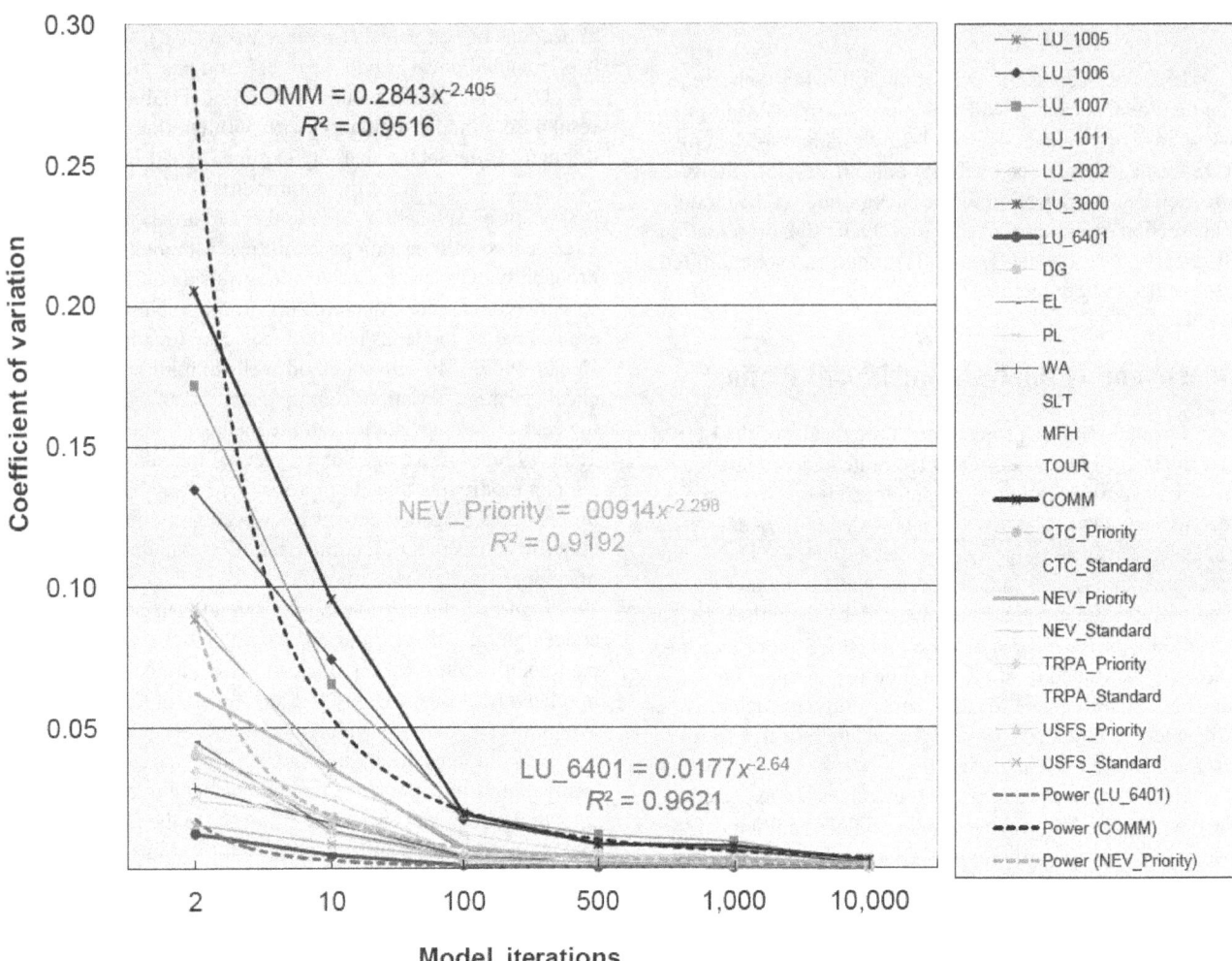

Figure 4. Coefficients of variation results for each commodity type and increasing iteration used in the Land Use Simulation Model (LUSM). Included in the figure are select power regression equations (for example, Power (LU_6401) and their coefficient of determination (R2). Notations for parcels: LU_1005 is a parcel allocated to multifamily dwelling, two to four units; LU_1006 is a parcel allocated to multifamily dwelling, 5 to 10 units; LU_1007 is a parcel allocated to multifamily dwelling, 10 or more units; LU_1011 is a parcel allocated to single-family dwelling; LU_2002 is a parcel allocated to tourist accommodations; LU_3000 is a part of a parcel allocated to commercial floor area; LU_6401 is a parcel allocated to open space or retired; DG is a parcel is allocated to Douglas County; EL is a parcel is allocated to Eldorado County; PL is a parcel allocated to Placer County; WA is a parcel allocated to Washoe County; SLT is a parcel allocated to the City of South Lake Tahoe; MFH is a parcel allocated to multifamily dwelling; TOUR is a parcel allocated to tourist accommodations; COMM is a part of a parcel allocated to commercial area; CTC_Priority is a sensitive parcel allocated to the retirement program of the California Tahoe Conservancy; CTC_Standard is a nonsensitive parcel allocated to the retirement program of the California Tahoe Conservancy; NEV_Priority is a sensitive parcel allocated to the retirement program of the Nevada Division of State Lands; NEV_Standard is a nonsensitive parcel allocated to the retirement program of the Nevada State Lands; TRPA_Priority is a sensitive parcel allocated to the retirement program of the Tahoe Regional Planning Agency (TRPA); TRPA_standard is a nonsensitive parcel allocated to the retirement program of the TRPA; USFS_Priority is a sensitive parcel allocated to the retirement program of the U.S. Forest Service (USFS); USFS_Standard is a nonsensitive parcel allocated to the retirement program of the USFS.

Table 16. Coefficients of variation (CV) by allocation type used in the Land Use Simulation Model.

[MFD and LU_1005, is a parcel is allocated to multifamily dwelling, 2 to 4 units; MFD and LU_1006, is a parcel is allocated to multifamily dwelling, 5 to 10 units; MFD and LU_1007, is a parcel is allocated to multifamily dwelling, 10 or more units; SFD and LU_1011, is a parcel is allocated to single-family dwelling; TAU and LU_2002, is a parcel is allocated to tourist accommodations; LU_3000 is a part of a parcel allocated to commercial area; LU_6401 is a parcel allocated to open space or retired; DG is a parcel allocated to Douglas County; EL is a parcel allocated to Eldorado County; PL is a parcel allocated to Placer County; WA is a parcel allocated to Washoe County; SLT is a parcel allocated to the City of South Lake Tahoe; CTC_Priority is a sensitive parcel allocated to the retirement program of the California Tahoe Conservancy (CTC); CTC_Standard is a nonsensitive parcel allocated to the retirement program of the CTC; NEV_Priority is a sensitive parcel allocated to the retirement program of the Nevada Division of State Lands; NEV_Standard is a nonsensitive parcel allocated to the retirement program of the Nevada State Lands; TRPA_Priority is a sensitive parcel allocated to the retirement program of the Tahoe Regional Planning Agency (TRPA); TRPA_standard is a nonsensitive parcel allocated to the retirement program of the TRPA; USFS_Priority is a sensitive parcel allocated to the retirement program of the U.S. Forest Service (USFS); USFS_Standard is a non-sensitive parcel allocated to the retirement program of the USFS]

	Iterations/CV precision					
Allocation type	2	10	100	500	1,000	10,000
MFD (2 to 4) (LU_1005)	0.0888	0.0361	0.0062	0.0045	0.0040	0.0010
MFD (5 to 10) (LU_1006)	0.1347	0.0743	0.0173	0.0087	0.0065	0.0013
MFD (10+) (LU_1007)	0.1716	0.0653	0.0189	0.0117	0.0095	0.0017
SFD (LU_1011)	0.0254	0.0118	0.0023	0.0011	0.0006	0.0002
TAU (LU_2002)	0.0930	0.0304	0.0120	0.0048	0.0038	0.0014
Commercial (LU_3000)	0.2053	0.0958	0.0194	0.0085	0.0078	0.0028
Retired (LU_6401)	0.0124	0.0048	0.0010	0.0005	0.0003	0.0001
DG	0.0241	0.0189	0.0057	0.0026	0.0014	0.0004
EL	0.0454	0.0133	0.0040	0.0022	0.0011	0.0004
PL	0.0383	0.0174	0.0050	0.0015	0.0019	0.0005
WA	0.0285	0.0159	0.0045	0.0022	0.0012	0.0004
SLT	0.0378	0.0217	0.0053	0.0025	0.0019	0.0005
MFH	0.0727	0.0287	0.0054	0.0034	0.0031	0.0007
TOUR	0.0930	0.0304	0.0120	0.0048	0.0038	0.0014
COMM	0.2053	0.0958	0.0194	0.0085	0.0078	0.0028
CTC_Priority	0.0401	0.0126	0.0031	0.0016	0.0010	0.0004
CTC_Standard	0.0322	0.0115	0.0018	0.0020	0.0011	0.0004
NEV_Priority	0.0623	0.0354	0.0075	0.0028	0.0035	0.0010
NEV_Standard	0.0407	0.0241	0.0045	0.0034	0.0014	0.0006
TRPA_Priority	0.0348	0.0184	0.0050	0.0038	0.0020	0.0006
TRPA_Standard	0.0233	0.0076	0.0028	0.0025	0.0011	0.0004
USFS_Priority	0.0422	0.0167	0.0026	0.0021	0.0013	0.0005
USFS_Standard	0.0155	0.0087	0.0035	0.0021	0.0010	0.0003

Table 17. Annual number of transitions by land-use type as produced by the Land Use Simulation Model.

[MFD, multifamily dwelling; SFD, single-family dwelling; TAU, tourist accommodation unit; -, represent null values]

Description	Years																				Totals
	Y1	Y2	Y3	Y4	Y5	Y6	Y7	Y8	Y9	Y10	Y11	Y12	Y13	Y14	Y15	Y16	Y17	Y18	Y19	Y20	
Vacant	0	0	0	0	0	0	0	0	0	0	0	0	0	0	0	0	0	0	0	0	-
MFD (2–4)	13	9	12	9	16	9	8	6	8	12	10	4	0	0	0	0	0	0	0	0	114
MFD (5–10)	1	2	2	1	2	2	2	2	2	1	3	1	1	0	0	0	0	0	0	0	16
MFD (10+)	1	1	0	1	0	1	1	1	1	1	0	1	1	0	0	0	0	0	0	0	7
SFD	187	203	173	173	128	155	123	166	163	143	149	89	0	0	0	0	0	0	0	0	1,849
TAU	2	2	3	2	3	4	7	3	3	2	3	1	0	0	0	0	0	0	0	0	31
Commercial	3	4	2	2	0	0	0	10	3	3	2	1	1	0	0	0	0	0	0	0	29
Retired	357	368	407	379	335	389	375	347	362	357	308	254	0	0	0	0	0	0	0	0	4,236
Totals	562	587	597	566	483	559	515	534	540	518	473	349	0	0	0	0	0	0	0	0	0

Table 18. Annual pool size of remaining vacant parcels available for transition as produced by the Land Use Simulation Model.

[MFD, multifamily dwelling; SFD, single-family dwelling; TAU, tourist accommodation unit; CTC, California Tahoe Conservancy; NDSL, Nevada Division of State Lands; TRPA, Tahoe Regional Planning Agency; USFS, U.S. Forest Service; SLT, City of South Lake Tahoe; Co., County]

Description	Y0	Y1	Y2	Y3	Y4	Y5	Y6	Y7	Y8	Y9	Y10	Y11	Y12	Y13	Y14	Y15	Y16	Y17	Y18	Y19	Y20	Totals
Douglas Co. parcels	150	123	89	63	39	16	2	0	0	0	0	0	0	0	0	0	0	0	0	0	0	480
El Dorado Co. parcels	1,823	1,649	1,475	1,305	1,147	1,019	849	697	533	341	180	45	0	0	0	0	0	0	0	0	0	11,061
Placer Co. Parcels	1,130	1,109	1,014	913	813	731	625	535	421	295	168	66	0	0	0	0	0	0	0	0	0	7,817
Placer Co. parcels—sensitive	60	0	0	0	0	0	0	0	0	0	0	0	0	0	0	0	0	0	0	0	0	60
Washoe Co. parcels	155	105	52	22	1	0	0	0	0	0	0	0	0	0	0	0	0	0	0	0	0	335
City of SLT parcels	1,268	1,177	1,089	991	888	801	707	600	487	362	234	107	0	0	0	0	0	0	0	0	0	8,708
MFD parcels	797	716	638	561	498	431	358	300	236	164	86	29	0	0	0	0	0	0	0	0	0	4,811
TAU parcels	98	91	83	70	62	55	47	34	26	18	12	4	0	0	0	0	0	0	0	0	0	598
Commerical parcels	286	263	241	222	203	184	166	146	121	91	61	31	0	0	0	0	0	0	0	0	0	2,012
CTC priority pe-tirement parcels	1,299	1,145	978	791	616	459	266	106	0	0	0	0	0	0	0	0	0	0	0	0	0	5,658
CTC secondary retirement parcels	4,463	4,155	3,846	3,521	3,209	2,939	2,621	2,299	1,880	1,340	822	349	0	0	0	0	0	0	0	0	0	31,441
NDSL priority retirement parcels	260	212	154	110	75	47	24	6	0	0	0	0	0	0	0	0	0	0	0	0	0	886
NDSL secondary retirement Parcels	259	208	155	115	70	43	18	3	0	0	0	0	0	0	0	0	0	0	0	0	0	870
TRPA priority retirement parcels	1,559	1,357	1,131	901	691	505	290	111	0	0	0	0	0	0	0	0	0	0	0	0	0	6,544
TRPA secondary retirement Parcels	4,722	4,362	4,001	3,635	3,279	2,982	2,639	2,302	1,880	1,340	822	349	0	0	0	0	0	0	0	0	0	32,311
USFS priority retirement parcels	1,559	1,357	1,131	901	691	505	290	111	0	0	0	0	0	0	0	0	0	0	0	0	0	6,544
USFS secondary retirement parcels	4,722	4,001	4,001	3,635	3,279	2,982	2,639	2,302	1,880	1,340	822	349	0	0	0	0	0	0	0	0	0	32,311
Totals	24,610	20,075	20,075	17,751	15,558	13,693	11,537	9,549	7,461	5,288	3,205	1,328	0	0	0	0	0	0	0	0	0	0

Table 19. Summary of vacant parcels entering into a particular commodity with three levels of probability as produced by the Land Use Simulation Model for the Lake Tahoe Basin.

Commodity type	Probability class	Number of parcels
Open space/retired	Low	3,235
	Medium	1,407
	High	1,639
Subtotal		6,281
Single-family dwelling	Low	1,770
	Medium	2,575
	High	241
Subtotal		4,586
Multifamily dwelling (2–4 units)	Low	72
	Medium	319
	High	259
Subtotal		650
Multifamily dwelling (5–10 units)	Low	47
	Medium	40
	High	12
Subtotal		97
Multifamily dwelling (10+ units)	Low	18
	Medium	21
	High	9
Subtotal		48
Tourist accommodation units	Low	17
	Medium	23
	High	58
Subtotal		108
Commercial	Low	100
	Medium	63
	High	121
Subtotal		284
Total		12,054

Discussion

The results suggest a number of interesting and insightful observations. In table 16, the CV results behaved as expected. Each commodity's CV decreased with increasing iterations of the model. Of the 138 values displayed in table 16, only two deviate from the downward trend. Figure 4 indicates the same decreasing trend across all commodities and allocation types. Furthermore, commercial commodities show the largest decrease, Nevada priority commodities show a mid-range decrease, and the retired allocation shows the smallest decreasing trend. Note that all three of the coefficient of determination (R^2) values of the power regression equations explain more than 90 percent of the variations in the data. The analysis of these results indicates an overall increase in the model's precision as a result of increased iterations.

As shown in table 17 and figures 5 and 6, most of the commodities are transitioned to SFD and retired.[18] Given the default values of the model (table 12), this makes intuitive sense, as the allocation demand values are greatest for SFDs in the counties and retirement in the programs. Furthermore, no transitions enter into *vacant*, as vacant parcels are a source of allocations, not a sink (table 17). Except for the case where lots have split, this also makes intuitive sense and is corroborated by the land-management practices in the basin. Also, of all the MFD units (2 to 4, 5 to 10, and 10 plus), the most frequently allocated is the 2–4 unit size, which makes sense as one would need larger parcels—which are in shorter supply in the basin, as shown in the mean area of table 11—to add additional units. Finally, with respect to table 17, it is important to note that under the current regional plan, parcel designations, and initial conditions of the model, all transitions will occur by the 12th year of the simulation, thereby reaching the point where no more vacant parcels are left in the basin and the region is built out. This is an important scenario for TRPA to consider as they are working on their Regional Plan Update and environmental documentation.

Table 18 shows the allocations of the sensitive lots (as defined by the current data's BLCS derivation of the IPES score) in Placer County (60) are used up in the first year, which suggests that the second year's activities of the retirement programs were sufficient to satisfy the Vacant Lot Equation, as no more sensitive candidate parcels remain in Placer County. Although these specific numbers are most likely inaccurate because of database errors, it does demonstrate the model's ability to capture the regulatory mechanism of the Vacant Lot Equation. The results demonstrate that development actions and retirement actions are related; however,

the retirement agencies act independently of each other and of real-estate developers. Table 18 indicates that TRPA and USFS, whose jurisdictions cover the entire basin, have the same number of parcels available to them at the beginning of the model's execution. Furthermore, the total number of *Priority* and *Secondary* parcels available to NDSL indicates that 519 parcels are initially available in Nevada for retirement. Although all vacant parcels (6,281) are available to the retirement agencies, the total number of parcels available to be allocated to a development fate in the beginning of the model is only 5,767, suggesting that the remaining 514 are initially deemed unavailable because of their low BLCS and IPES scores. These results depend on the accuracy of the BLCS and IPES scores that were spatially formatted by USGS and provided by TRPA, respectively. Finally, regarding the transitions available to TRPA and USFS, initially it may seem curious that their values are identical over time. This is, however, reasonable and consistent with what one would expect as (1) the parcels available to them would be equivalent because of their collocated jurisdictions, and (2) over time, as the fates of various parcels are determined and they are taken out of the vacant pool, the remaining parcels would still be available to both agencies for retirement in any given year. The fact that they have different initial default values is relevant to the allocations that are being or have been made, not what remains to be made. As a supporting observation, it is interesting to note once all the parcels that are available to the NDSL have been exhausted in year seven, then the only retirement parcels left are in California. Thus starting in year eight, the parcels available to the CTC are equivalent to TRPA's and USFS's parcels.

Using the default values of the LUSM, table 19 indicates that the greatest number of parcels enter into open space/retired, followed in decreasing order by SFD, MFD 2 to 4, commercial, TAU, MFD 5 to 10, and finally MFD 10+. It is important to note that the total number of possible transitions (12,577) is greater than the total number of vacant parcels input to the LUSM (6,281). This shows there can be more than one possible fate—transition type or allocation—for any particular parcel. A single parcel can have multiple fates (in other words, vacant to both SFD and MFD). Furthermore, each fate has a certain probability assigned to it (in other words, 30 percent chance that it will transition to MFD).

In discussing figures 5 through 11, it appears the results make intuitive sense. The spatially constrained nature of the model by PAS placed commodity allocations in relevant and conceivable places. Given the size constraint on vacant parcels that could enter into retirement, many of the larger parcels are farther from the lake and on the periphery of urban areas. This mimics typical priorities of retirement agencies as they tend to focus on larger parcels that have a lower cost and are less "useful" in terms of development. Parcels that are contiguous to existing developed land are likely to cost more and be more attractive to developers. Given the typical desire of developers to be closer to the lake for recreation and aesthetic purposes, as well as to closer to commercial corridors and urban centers for basic service provision, it is reasonable to expect the

[18]Note that the summation of the number of parcels in the explanations of figures 5 through 11 do not equal exactly the totals in table 17 because they are the results of different model runs, where the number of iterations was different. Furthermore, it's a stochastic model, so each run with the same input values provides slightly different output results.

Table 20. Land-cover impacts in the Lake Tahoe Basin resulting from the Land Use Simulation Model's default scenario's land-use transitions by commodity type.

[Note: the "Water" land cover classes are an artifact primarily of the spatial join technique where parcels overlapped water, and secondarily where LULC dataset was misclassified during processing; ha., hectares; lodgepole pines (*Pinus contorta*), white and red fir (*Abies concolor* and *Abies magnifica*), quaking aspen (*Populus tremuloides*), willow (*Salix* spp.), Jeffery pine (*Pinus jeffreyi*), and Great Basin sagebrush (*Artemisia tridentate*), greenleaf manzanita (*Arctostaphylos patula*), huckleberry oak (*Quercus vacciniifolia*), mountain whitethorn (*Ceanothus cordulatus*), and whitebark pine (*Pinus albicaulis*)]

Existing land-cover type	Total area in basin (ha)	Count of parcels	Median size of parcels (ha)	Parcel size, standard deviation (ha)
Single-family dwelling				
Jeffrey pine	34.09	126	0.15	0.61
Basin sagebrush	15.76	51	0.13	1.00
Upper montane mixed shrub	14.43	7	0.25	4.75
Red fir	3.66	33	0.10	0.06
Water	2.71	5	0.52	0.34
Mixed conifer-fir	1.74	11	0.12	0.13
Perennial grasses/forbs	0.40	2	0.20	0.04
Huckleberry oak	0.34	5	0.07	0.02
Greenleaf manzanita	0.09	1		
Subtotals	73.23	241	0.14	0.23
Multifamily dwelling (2–4 units)				
Jeffrey pine	9.04	145	0.06	0.02
Basin sagebrush	3.68	62	0.06	0.02
White fir	1.69	26	0.06	0.01
Upper montane mixed shrub	1.04	19	0.05	0.01
Mixed conifer-fir	0.33	5	0.05	0.03
Greenleaf manzanita	0.10	1		
Quaking aspen	0.10	1		
Subtotals	15.96	259	0.06	0.02
Multifamily dwelling (5–10 units)				
Jeffrey pine	1.67	8	0.20	0.08
Basin sagebrush	0.93	4	0.24	0.02
Subtotals	2.60	12	0.22	0.05
Multifamily dwelling (10+ units)				
Jeffrey pine	7.99	5	1.62	0.99
Basin sagebrush	2.09	4	0.41	0.31
Subtotals	10.08	9	1.02	0.65
Tourist accommodation units				
Basin sagebrush	7.54	22	0.17	0.38
White fir	5.33	11	0.41	0.42
Jeffrey pine	4.21	20	0.11	0.22
Mixed conifer-fir	0.20	2	0.10	0.03
Upper montane mixed shrub	0.15	2	0.07	0.04
Unclassified	0.12	1		
Subtotals	17.55	58	0.11	0.22

Table 20. Land-cover impacts in the Lake Tahoe Basin resulting from the Land Use Simulation Model's default scenario's land-use transitions by commodity type.—Continued

[Note: the "Water" land cover classes are an artifact primarily of the spatial join technique where parcels overlapped water, and secondarily where LULC dataset was misclassified during processing; ha., hectares; lodgepole pines (*Pinus contorta*), white and red fir (*Abies concolor* and *Abies magnifica*), quaking aspen (*Populus tremuloides*), willow (*Salix* spp.), Jeffery pine (*Pinus jeffreyi*), and Great Basin sagebrush (*Artemisia tridentate),* greenleaf manzanita (*Arctostaphylos patula*), huckleberry oak (*Quercus vacciniifolia*), mountain whitethorn (*Ceanothus cordulatus*), and whitebark pine (*Pinus albicaulis*)]

Existing land-cover type	Total area in basin (ha)	Count of parcels	Median size of parcels (ha)	Parcel size, standard deviation (ha)
Commercial				
Basin sagebrush	4.54	69	0.05	0.07
Jeffrey pine	2.10	35	0.05	0.06
Greenleaf manzanita	0.51	3	0.19	0.03
White fir	0.33	7	0.05	0.02
Mixed conifer-fir	0.33	4	0.07	0.04
Upper montane mixed shrub	0.22	2	0.11	0.08
Huckleberry oak	0.02	1		
Subtotals	8.04	121	0.06	0.05
Open space/retired				
Jeffrey pine	683.17	741	0.03	7.99
Basin sagebrush	351.59	329	0.07	5.29
White fir	283.72	180	0.06	6.61
Red fir	129.07	29	1.36	6.06
Huckleberry oak	82.84	61	0.03	5.87
Upper montane mixed shrub	78.30	61	0.08	4.16
Greenleaf manzanita	76.08	40	0.19	4.63
Quaking aspen	50.61	22	0.37	4.65
Lodgepole pine	48.84	5	0.15	21.50
Mixed conifer-fir	28.28	32	0.09	3.67
Perennial grasses/forbs	28.21	23	0.08	3.01
Whitebark pine	15.42	1		
Water	13.97	66	0.07	0.56
Willow	12.14	34	0.06	0.99
Barren	9.30	5	0.14	3.66
Unclassified	6.43	2	3.21	3.06
Unknown conifer	1.52	5	0.43	0.27
Mountain whitethorn	1.36	1		
Unknown shrub	0.23	2	0.12	0.01
Subtotals	1901.06	1,639	0.09	4.16
Totals for all land covers	2028.51	2,339	0.11	0.31

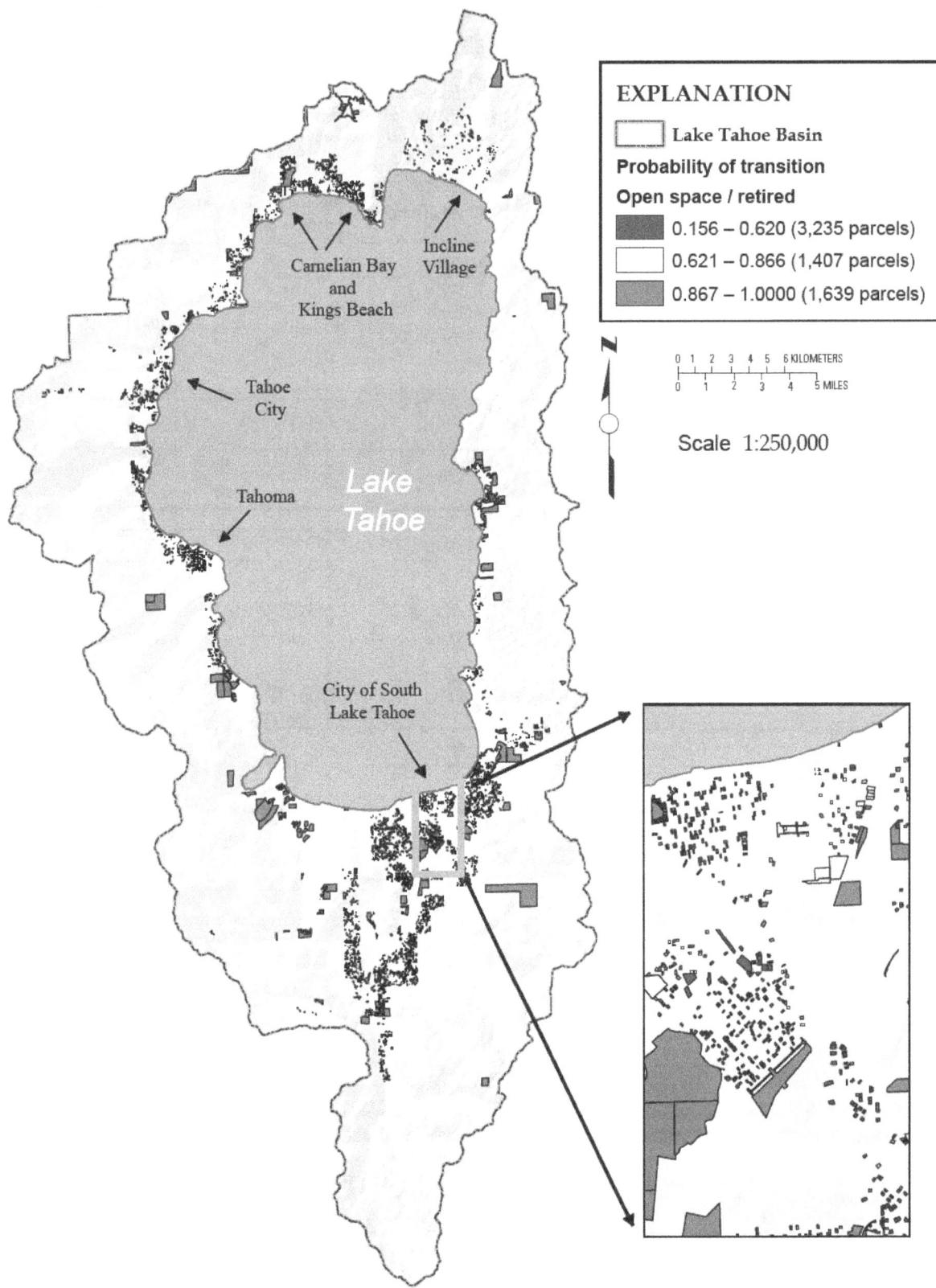

Figure 5. Map showing output of the Land Use Simulation Model (LUSM) for the Lake Tahoe Basin of vacant parcels transitioning to open space or retired parcel allocations. The output includes the ranges of probability of transition and the number of parcels in each range of probabilities. Source for shaded-relief model base map is U.S. Geological Survey National Elevation Dataset.

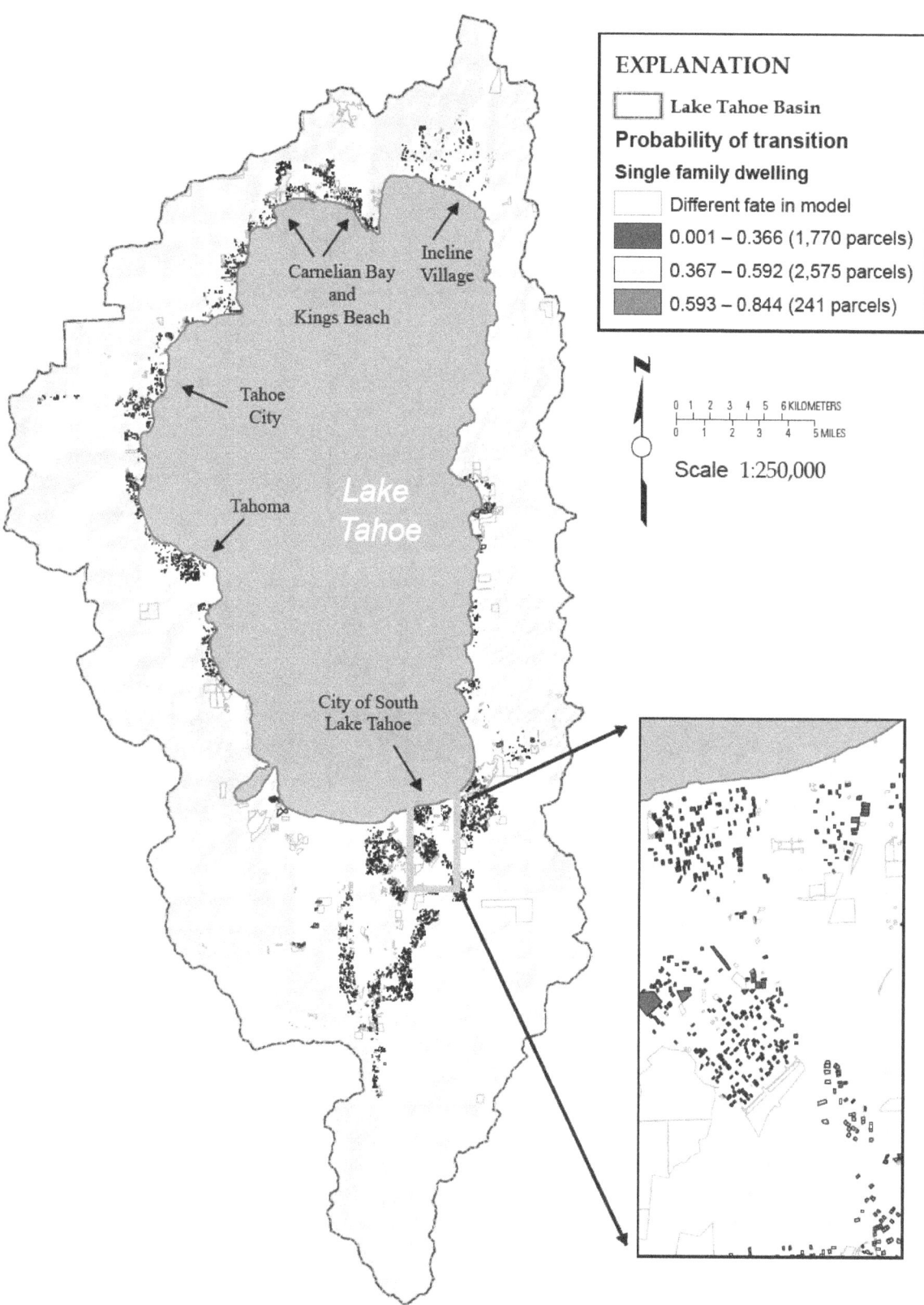

Figure 6. Map showing output of the Land Use Simulation Model (LUSM) for the Lake Tahoe Basin of vacant parcels transitioning to single-family dwelling allocations. The output includes the ranges of probability of transition and the number of parcels in each range of probabilities. Source for shaded-relief model base map is U.S. Geological Survey National Elevation Dataset.

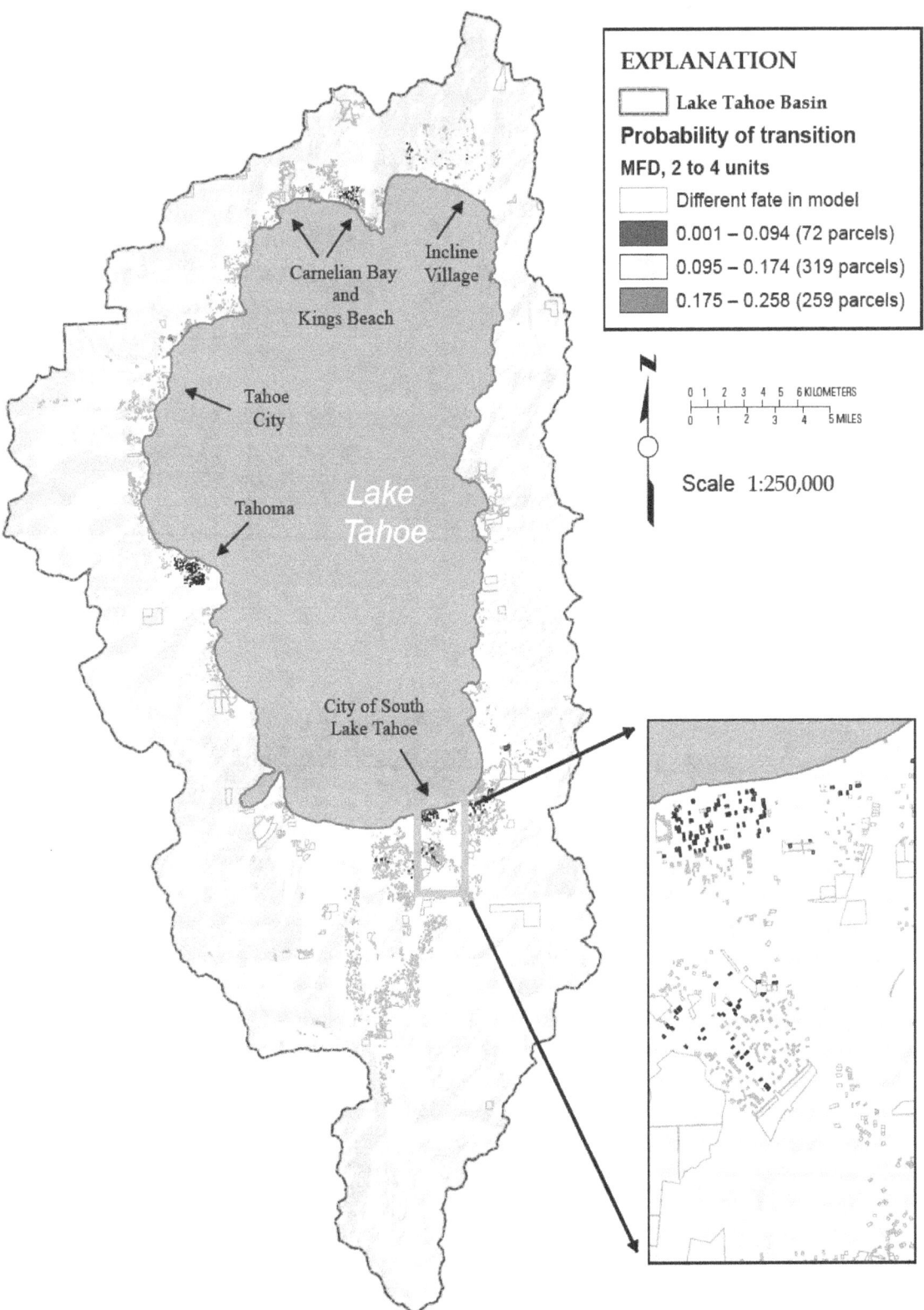

Figure 7. Map showing output of the Land Use Simulation Model (LUSM) for the Lake Tahoe Basin of vacant parcels transitioning to multifamily dwelling (MFD), 2 to 4 unit allocations. The output includes the ranges of probability of transition and the number of parcels in each range of probabilities. Source for shaded-relief model base map is U.S. Geological Survey National Elevation Dataset.

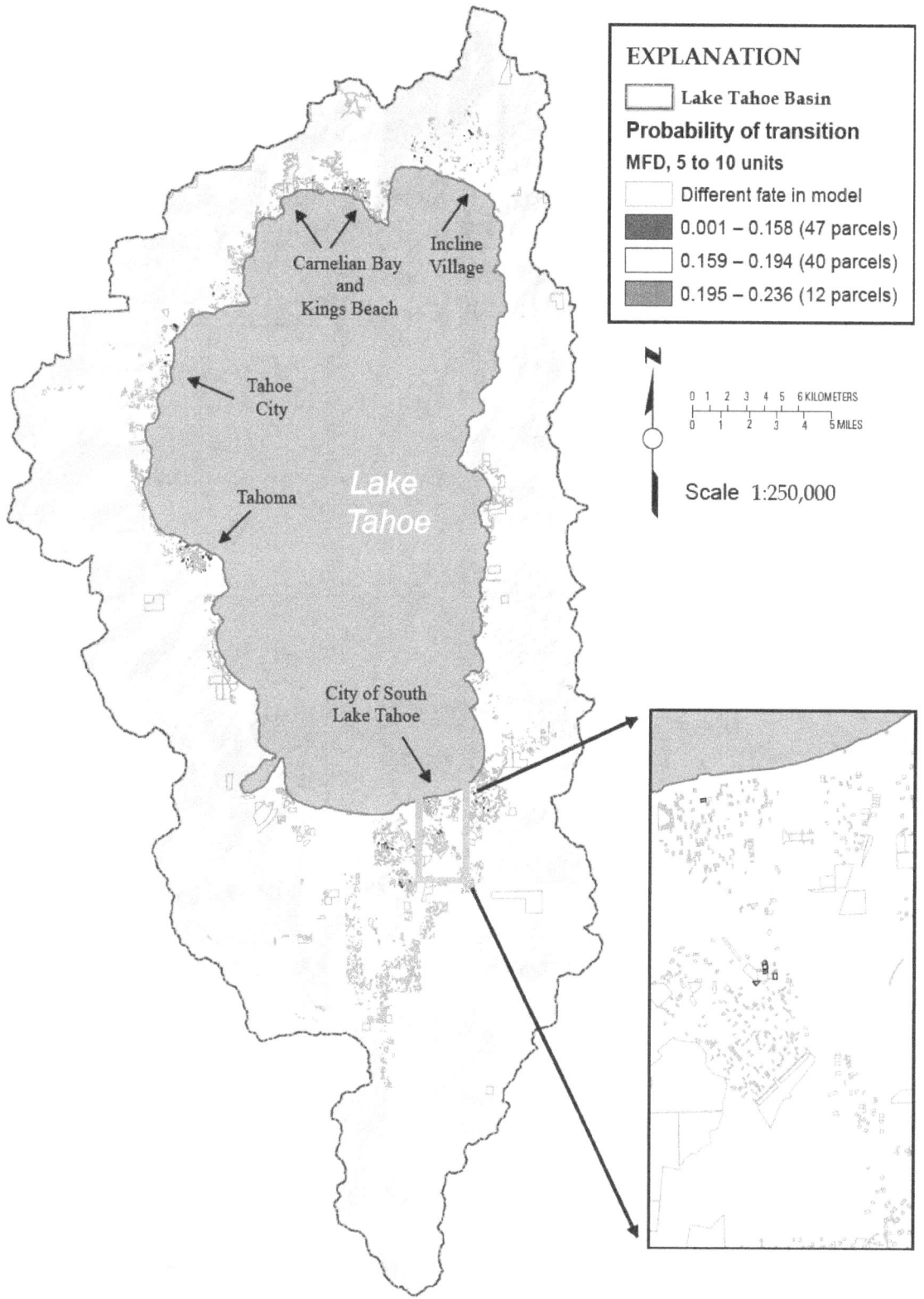

Figure 8. Map showing output of the Land Use Simulation Model (LUSM) for the Lake Tahoe Basin of vacant parcels transitioning to multifamily dwelling (MFD), 5 to 10 unit allocations. The output includes the ranges of probability of transition and the number of parcels in each range of probabilities. Source for shaded-relief model base map is U.S. Geological Survey National Elevation Dataset.

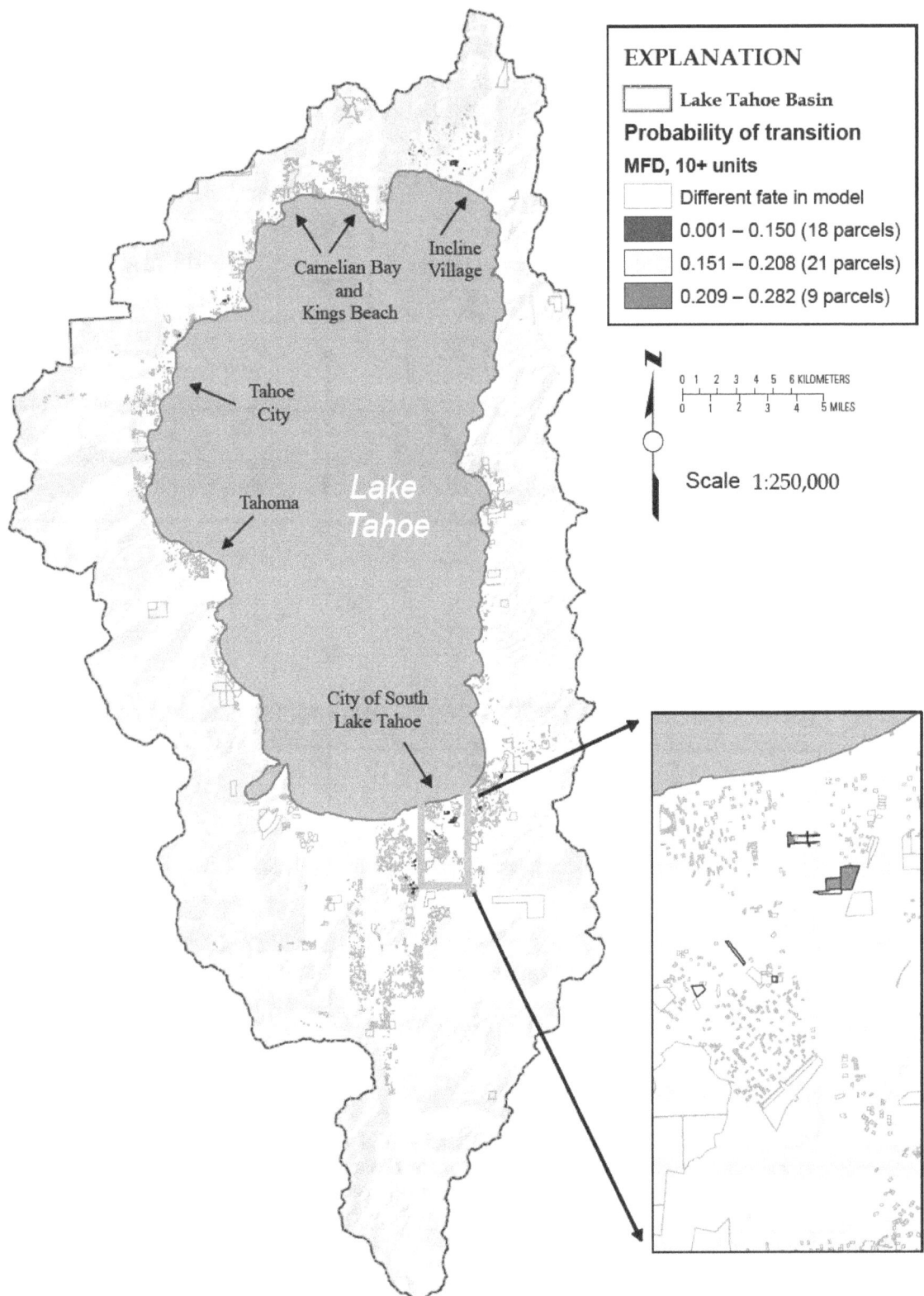

Figure 9. Map showing output of the Land Use Simulation Model (LUSM) for the Lake Tahoe Basin of vacant parcels transitioning to multifamily dwelling (MFD), 10+ unit allocations. The output includes the ranges of probability of transition and the number of parcels in each range of probabilities. Source for shaded-relief model base map is U.S. Geological Survey National Elevation Dataset.

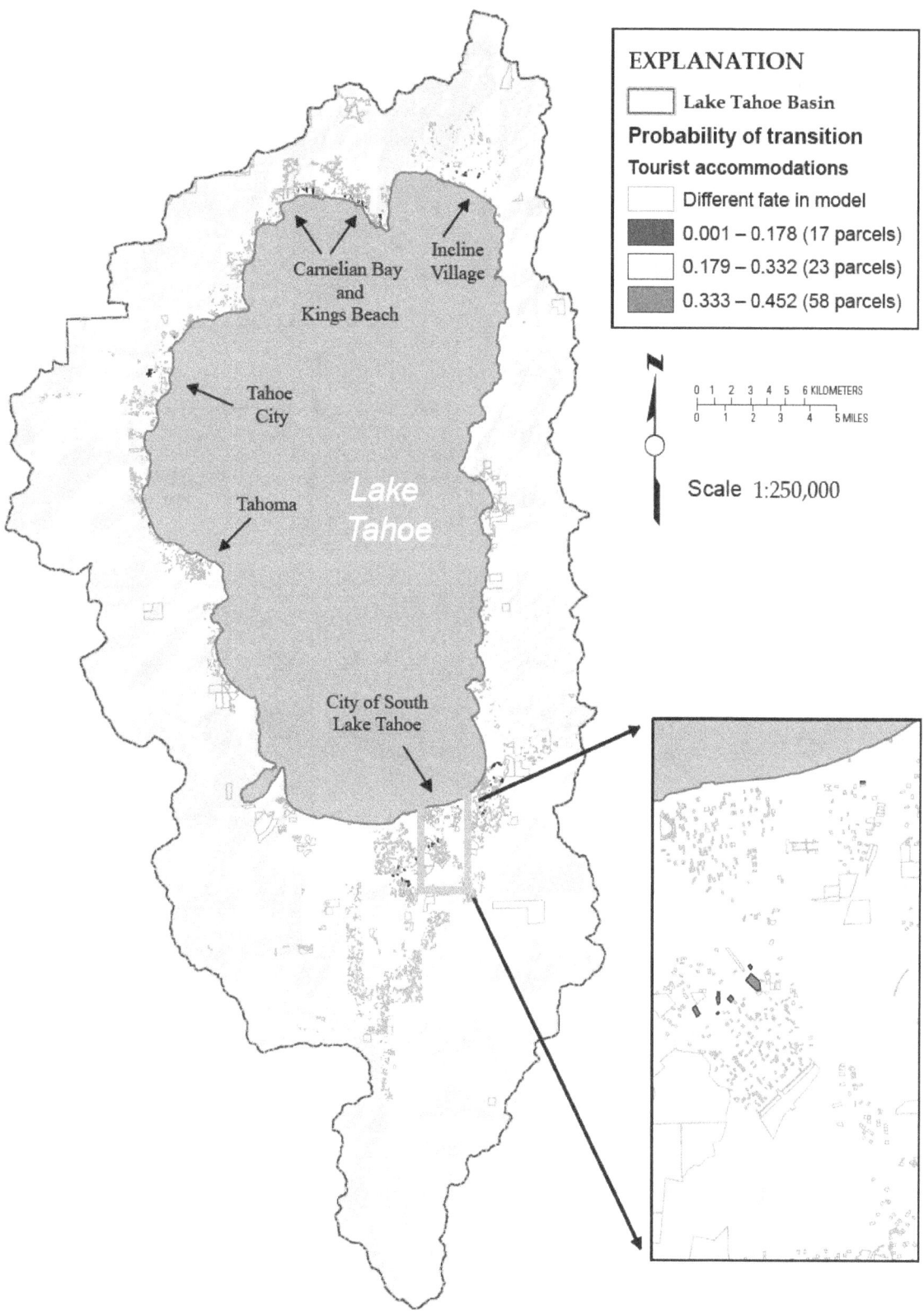

Figure 10. Map showing output of the Land Use Simulation Model (LUSM) for the Lake Tahoe Basin of vacant parcels transitioning to tourist accommodation allocations. The output includes the ranges of probability of transition and the number of parcels in each range of probabilities. Source for shaded-relief model base map is U.S. Geological Survey National Elevation Dataset.

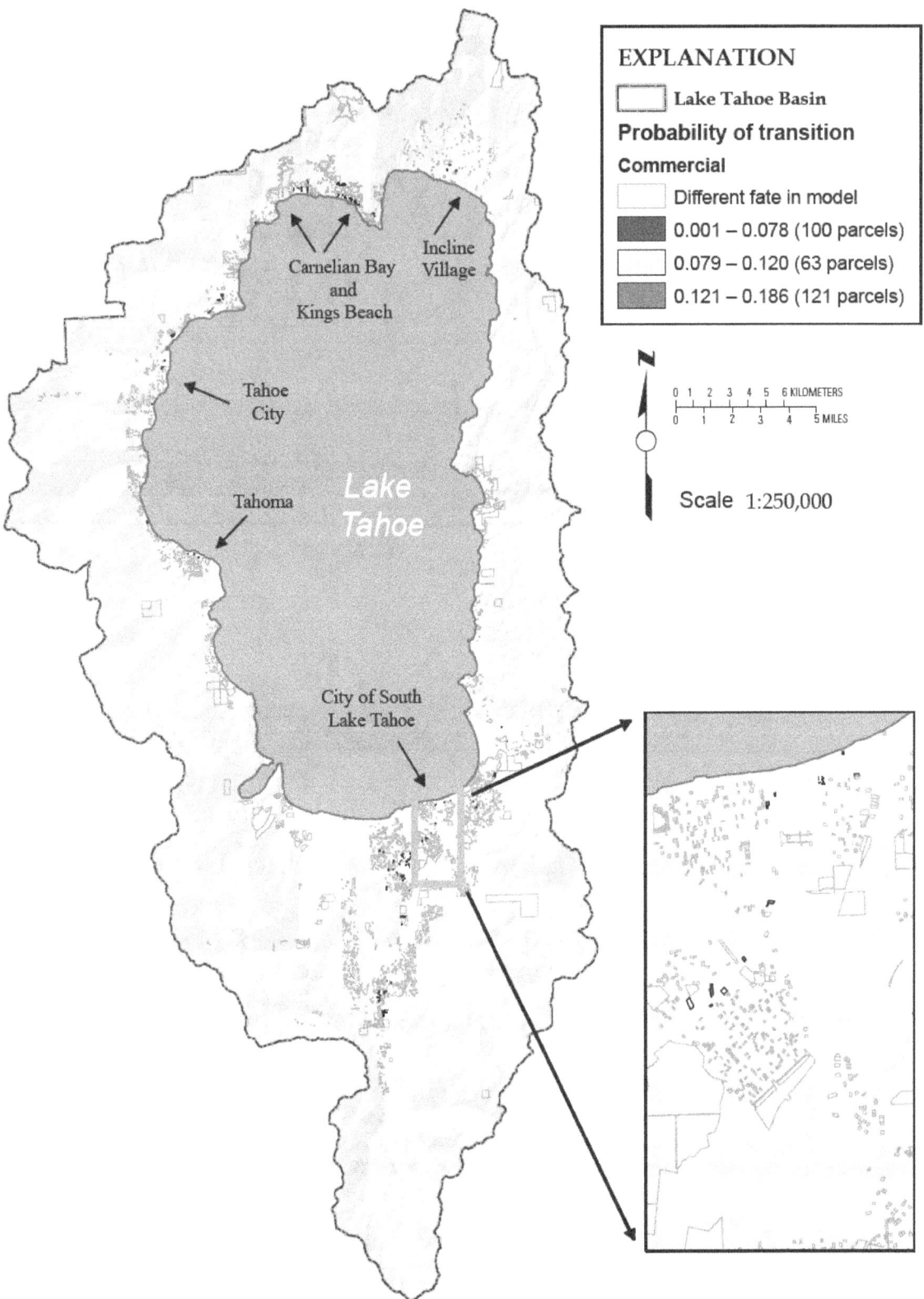

Figure 11. Map showing output of the Land Use Simulation Model (LUSM) for the Lake Tahoe Basin of vacant parcels transitioning to commercial allocations. The output includes the ranges of probability of transition and the number of parcels in each range of probabilities. Source for shaded-relief model base map is U.S. Geological Survey National Elevation Dataset.

spatial distribution of SFDs in the LUSM and their allocation to smaller lots. The TRPA density restrictions allow for one home on one lot—with certain setback and land capability restrictions—so the smaller lots can support SFDs. The density requirements of MFDs by PAS, however, are more restrictive because the number of units allowed is designated on a per acre basis, thereby requiring the availability of larger parcels. Furthermore, larger MFDs require a comparable number of allocations to offset their potential use elsewhere, which reduces the frequency of a large MFD being allocated. This applies to the area-restricted density of TAUs and commercial allocations as well—they need larger parcels. As for their location, the MFD, TAU, and commercial allocations are situated intuitively and as would be expected in and around the larger towns of the Lake Tahoe Basin.

As shown in table 20, the linkage to another spatially explicit dataset focuses on the future impacts of the default LUSM scenario on land cover. Highly dependent on the quality and accuracy of the remote-sensing classification of IKONOS imagery, the analysis indicates the high degree of impact on Jeffery pine and sagebrush, as those two land-cover classes have the two greatest areas across all development commodities. In terms of open space and retired parcels, Jeffrey pine and sagebrush were also the greatest two areas. Furthermore, more than 50 ha of quaking aspen was preserved—which is an important habitat for Tahoe wildlife (Manley and others, 2007)—and the median size parcel of red fir—which is known to occur at higher elevations—suggests their location may be outside of urban, developable areas. The median parcel size of retired commodities (0.09 ha) is notably smaller as compared to the other commodities (say, SFD of 0.14 ha). As was expected, the larger the MFD unit, the more land that was required to meet the density requirements (in other words, increasing median sizes of 0.06, 0.22, and 1.02). Given the desire of resource agencies to improve management of urban stormwater, the urban wildfire interface, and urban biodiversity, this is just one example of linking an existing, ancillary spatial dataset to one scenario of the LUSM to provide insight into how the tool could be used to consider priorities of resource managers, the configurations of the landscape, and the analysis of repercussions of the transitions of particular parcels.

Limitations

The LUSM is a simulation tool meant for planning purposes. The default values were derived mostly from historical activity, which may not be entirely pertinent to future activities. Also, the model resolves to the parcel scale. Subparcel considerations are not currently handled in the model (for example, slope, aspect, soils, BMPs, vegetation characteristics). Furthermore, it resolves to only a certain land-use class, mostly to the first level of TRPA's hierarchical scheme. The model is able to represent the basic commercial level but not the mixed-use development patterns being considered in the TRPA Regional Plan Update. The first-order commercial

representation, however, is important to the regional plan updating process and the development of baseline and project alternatives for Environmental Impact Reports and Environmental Impact Statements.

One specific limitation of the model is related to residential allocations. As summarized in the Residential Allocation Performance table (table 13), if the LUSM were to perfectly mimic the realities of how planning works, the user ought to be able to change the values from the initial conditions on an annual basis. During each year, each jurisdiction is assessed in accordance with seven criteria for allocation deductions or enhancements. This level of detail is too difficult to model accurately, as is the user's ability to determine how a particular jurisdiction will behave from year to year. Any model of policy effects, however, inherently includes assumptions about social behavior (Yearley, 1999). For example, the values in table 8 indicate the variation in historic allocations, which generally are greater than the maximum default value for each jurisdiction. As such, the current and past policies—which were used to guide the development of the LUSM's logic—are not in accord with the behavior of the planners at TRPA over a two-decade timespan. This discrepancy is typical when land-use planning models try to characterize and encapsulate the vagaries of human behavior and decisionmaking.

The artificial linkage between BLCS and IPES is useful for the purposes of the LUSM; however, the degree to which the BLCS can fill in IPES data gaps is suspect as BLCS is meant for nonresidential parcels and the translation between the two systems is purely artificial. The BLCS is an older land capability system. Furthermore, field verifications can be requested for specific parcels (Bernknopf and others, 2003), and are called "land capability verifications" (TRPA, 2008c) to better determine land capability. Given the likely flaws in the parcel layer provided by TRPA[19], the authors thought it necessary to create an alternate means for a comprehensive database that depended on existing TRPA Code of Ordinances and datasets (in other words, the BLCS) to compensate for data-limited parcels. The lack of IPES values in Placer County is particularly relevant to satisfying the Vacant Lot Equation. A section of the logic of the model was developed to preferentially select parcels with low IPES scores (in other words, priority) in an effort to take sensitive parcels out of rotation, then secondarily release the remaining residential parcels with scores below 726. The ability of the model to do this was extended to include all agency programs, which are active beyond Placer County, so that the agencies can select their preference for sensitive lots that have low IPES scores. For both situations, however, the execution of the model depends on—and is limited by—having adequate and comprehensive IPES data.

We considered the inclusion of redevelopment in the model. Using a sample of Douglas County Assessor data, five

---

[19]For example, none of the vacant parcels in Placer County had an IPES score.

metrics were derived for targeting potential areas of commercial redevelopment (Forney and others, 2008). In an attempt to provide an indication of whether a parcel is undervalued in relation to its value if it were redeveloped, the economic value-based metrics included both area and location. Given the difficulty of deciding on an appropriate metric, lack of relevant literature on the theory and application of redevelopment in land-use models and planning, the possibility of litigation[20], and the unavailability of consistent and comprehensive assessor data for the entire Lake Tahoe Basin, redevelopment was not included in the LUSM. The LUSM only considers parcels that are currently vacant.

Suggested Next Steps

This section describes the considerations for the next steps of the LUSM related to data management for TRPA, the model's maintenance and updates, and additional research avenues to pursue for applications and natural-resource management in the Lake Tahoe Basin.

Data Management for TRPA

Having worked extensively with the PAS and parcel data provided by TRPA, we can offer suggestions for upgrading these input datasets to improve the model's outputs and facilitate the continued utility of the LUSM. A record of the PAS Uses needs to be added and kept current with any changes that occur in the future (for example, redefining the boundaries or allowed uses). Furthermore, the linework of the PASs could be improved as some were observed to bisect county boundaries. Data management practices could be improved to better distinguish null values from other values of meaning, and the records of various written documents should be consistent with the digital records. The parcel layer needs a more accurate and comprehensive representation of IPES scores.

Currently, none of the parcel data used in the model for Placer County have IPES scores, so they need to be added, because IPES scores are more meaningful to a parcel than BLCS scores. Furthermore, for the parcels of the entire basin, it is unclear if an IPES value of zero represents the fact that the parcel is unbuildable or that it simply hasn't ever been scored. Making that distinction would be useful. Finally, tracking the TDR of a parcel, including both where it came from and where it ended up to create multifamily dwellings and increased urban density, would be useful. This is particularly important if the TDR moves from one watershed to another or from one jurisdiction to another. With these improved data, the existing version of the LUSM could be improved to better represent the current and planned management practices of TRPA in the basin.

Maintenance and Updates

TRPA has been empowered with ownership, training, and oversight of the LUSM. The model's underlying Python code has been provided to TRPA, and their GIS analysts can alter and adapt it. Given how the Servoy© interface calls the code, changes to it ought to be easily reflected in the Servoy© interface. TRPA maintains and manages the Servoy© licenses, and is Servoy©'s primary point of contact. If extensive modifications are desired, TRPA can contract with Servoy© to synchronize the back-end changes with the front-end user interface.

If the new zoning map associated with the proposed transect method (table 7) becomes a reality, a more extensive reconfiguring of the back-end of the model would be required. The standard model language that was used and the extensive comments on the model code (Forney and Oldham, 2011) should enable developers and programmers to easily make necessary revisions in the future. The implementation of the transect method and associated model improvements would likely include the projection of different land-use types and TRPA commodities into the future.

Additional Research Avenues

The existence, form, and outputs of the LUSM provide for additional avenues of research. These include:

- Conduct model runs for a variety of scenarios (in other words, parcel-retirement oriented, development oriented, special-use oriented, higher-density commercial and tourist zoning, and jurisdictional priorities) and analyze spatial patterns of particular commodities and their locations to each other and to other spatially explicit datasets.

- Derive additional techniques to determine the most likely fate for a particular parcel, such as normalizing the probabilities within a commodity and then comparing across commodity types.

[20]Citing the determination of a blighted area in need of redevelopment in New York City, the Urban Development Corporation (aka the Empire State Development Corporation) recently received a favorable—yet highly contentious—over turning of a lower court's decision to allow its use of eminent domain to build a new part of Columbia University's campus in Manhattanville (New York Times, Charles V. Bagli, "Court Upholds Columbia Campus Expansion Plan", June 23, 2010). The corporation, a public entity, used the findings of two private urban planning consulting firms to document the conditions of blight, which enabled the use of eminent domain. Although the reports are not readily available to the public, journalistic reports of the documents suggested that such factors as physical blight, crime, deteriorated structures, building-code violations, safety conditions, tenant vacancy, and unsanitary conditions were included. These factors are highly subjective, and do not have clear methods currently supported in the scientific literature. As such, the existing methods for targeting redevelopment in urban and blighted areas are unclear and poorly vetted, making them difficult to employ in land-use change modeling and planning.

- Conduct local sensitivity analyses to determine the predominant drivers of change in output as related to the default values of the model's initial conditions.

- Compare the model's future distributions of commodities and allocations with current and past distributions of commodities across the basin. This could include using the multitemporal analysis of cover type and change detection in the Trout Creek, Bijou, and Upper Truckee watersheds (Raumann and Cablk, 2008) to cross-reference the historical trends and locations of land-use change.

- Corroborate the LUSM with Verburg and others's (2002) Kappa statistic by obtaining a historic parcel dataset that is run through time to present day. Given a suitable, older parcel dataset to input to the LUSM, the model can be run forward to match a particular time step of available data. For that given year, cross-referencing the output of the model with the reality of what is on the landscape would be an effective means to corroborate the LUSM.

- Apply the raster-based CLUE-S model to the Lake Tahoe Basin and compare its results with the LUSM output to investigate the differences that may occur from including historical validation, empirical forecasting, and raster-based agents in land-use change modeling.

- Given the fact that the historical legacy of a parcel's land use is useful when considering its management and possible restoration, Raumann and Cablk's work (2008) could be referenced to interpret the feasibility of particular land-preservation scenarios provided by the LUSM's retirement output.

- In terms of fire susceptibility, severity of threat, and possible mitigation techniques, compare LUSM output to other spatially explicit fire-modeling outputs.

- In terms of water quality, the TMDL erosion potential analysis (Simon and others, 2003) could be cross-referenced with the relative erosion potential of the BLCS (table 3) and the relative erosion hazard of IPES (table 5) to address the erosion potential of different LUSM scenario outputs.

- In terms of urban biodiversity, improved scenarios could be provided to Manley and others (2007) as related to biodiversity potential and the trajectory of habitat availability. Factors that impact habitat condition at the neighborhood/landscape scale include amount of conifer vegetation, amount of aspen and riparian vegetation (especially for "habitat specialists" such as shrews (*Sorex* spp.)), and presence of development (Manley and others, 2007). This could improve the temporal aspects of spatially explicit population

models (Conroy and others, 1995; Dunning and others, 1995; Holt and others, 1995; McGarigal and McComb, 1995; Turner and others, 2001). The scenarios generated by the LUSM may provide a better tool for estimating the number and location of parcels that would be developed or remain open space, thereby providing a constrained, realistic approach to population viability and habitat connectivity in the face of an uncertain future.

- With the model's outputs, calculate estimates of population growth and human water demand. Compare the estimates with other estimates from demographers.

- If a good bare-earth model can be derived from the light detection and ranging (lidar) data of SNPLMA's Capital Improvement Program, slope, aspect, amount of imperviousness (derived from remote sensing imagery), hydrologic proximity to the lake, and other variables could be added. Related to improvements in watershed rainfall-runoff modeling efforts, incorporating better measures of impervious coverage by commodity class from distributions of remotely sensed observations would be useful to understanding future hydrologic conditions.

- Use the knowledge gained in developing the LUSM to apply it in a different geographic location where natural resources and planning for their sustainability depends on addressing scenarios of stochastic, restricted land-use/land-cover change.

Overall, the LUSM can provide multiple scenarios of future conditions of the landscape matrix, allowing for landscape trajectory analysis (Cushman and McGarigal, 2007) and its potential linkages and ramifications to other natural-resource issues. The natural-resource processes need to be either (1) assumed to be reasonably static over the timespan of the model's run (for example, soils—in certain locations—and geology over 20 years) or (2) modeled and have their state changes characterized into the future in accordance with the LUSM results. In relation to anthropogenic influences, the LUSM and natural-resource linkages could provide more meaningful distributions of development and open space, as well as insights into the ramifications of today's choices under current regulations on the future patterns of the landscape.

Summary and Conclusion

This report has described the capstone of the multiyear, TDSS of the WGSC, namely the LUSM. The LUSM is useful for simulating user-driven scenarios—with measures of uncertainty—of the possible state of the future. This manuscript presented the natural, anthropogenic, and management context related to the development of the LUSM in the Lake Tahoe Basin, the LUSM's data needs and quality, the details of the model's logic and decision rules, the communication and collaboration with TRPA, and the design criteria laid out in conjunction with TRPA. The LUSM tool has met the demands of (1) updating the model with new parcel data; (2) disaggregating it to an individual parcel level and maintaining the APN as the key data identifier; (3) removing the hard coding of the development intention and adding a spatially constrained, stochastic element to it; (4) adding the allowable uses of the PASs and the building density by zoning type; (5) filling in the gaps of missing IPES scores for all relevant parcels with BLCS scores; (6) outputting the results to a physical location on a map, and showing the feasibility of linking them to other geospatial analyses, models, and databases; (7) pursuing the feasibility of incorporating redevelopment to the model to assist with the TRPA Regional Plan Update; (8) producing probabilistic results related to the likelihood of a parcel's transition to a new land-use state; and (9) migrating the Servoy© database to another TRPA server for its long-term use and Web-based access. Furthermore, this report included relevant land-use change modeling theory, characterization of relevant organizational and institutional frameworks, and literature review, as well as detailed descriptions of the model's default values, relevant assumptions, model assessments, and results. This report continued to discuss the results, the LUSM's limitations and potential improvements, and suggestions for TRPA's data management, maintenance, and updates. Finally, the report concludes with consideration of additional, stand-alone research avenues that could provide additional insights to improve the management of natural resources in the Lake Tahoe Basin.

Acknowledgments

This research has been supported by the USGS Geographic Analysis and Monitoring Program and funded by the Southern Nevada Public Land Management Act. The research has been conducted in cooperation with the TRPA. The authors would like to thank Laura Dinitz, David Halsing, Dr. Richard Taketa, Mara Tongue, and Susan Benjamin for their thorough and insightful reviews, as well as WGSC staff that both supported and laid the groundwork of this effort such as Dr. Richard Bernknopf, David Halsing, Mark Hessenflow, Sean Devlin, Elizabeth Duffie, Peter Ng, Michael Gould, and Mara Tongue.

References

2ndNature, LLC, 2006, Lake Tahoe BMP monitoring evaluation process: synthesis of existing research: 2ndNature, LLC, Web site, accessed February, 2008, at http://www.2ndnaturellc.com/wp-content/uploads/2011/09/Final-Report_BMPSynthesis.pdf

Arnold, C.L., Jr., and Gibbons, C.J., 1996, Impervious surface coverage—The emergence of a key environmental indicator: Journal of the American Planning Association, v. 62, p. 243–258.

Bailey, R.G., 1974, Land-capability classification of the Lake Tahoe Basin, California-Nevada—A guide for planning: U.S. Department of Agriculture and the U.S. Forest Service, 44 p.

Balzter, H., Braun, P.W., and Köhler, W., 1998, Cellular automata models for vegetation dynamics: Ecological Modelling, v. 107, p. 113–25.

Berka, C., McCallum, D., and Wernick, B., 1995, Land use impacts on water quality—case studies in three watersheds: Presented at The Lower Fraser Basin in Transition—A Symposium and Workshop, May 4, 1995, Resource Management and Environmental Studies, University of British Columbia, Vancouver, Canada, p. 62–79.

Bernknopf, R.L., Champion, R., Duffie, E.S., Graffy, E.A., Ng, P., and Taketa, R., 2003, Final report to TRPA—Tahoe constrained optimization model: U.S. Geological Survey unpub. report submitted to the Tahoe Regional Planning Agency, 77 p.

Bolte, J.P., Hulse, D.W., Gregory, S.V., and Smith, C., 2006, Modeling biocomplexity—Actors, landscapes and alternative futures: Environmental Modelling and Software, v. 22, p. 570–579.

Boothe, D.B., 1991, Urbanization and the natural drainage system impact, solutions and prognosis: Northwest Environment Journal, v. 7, p. 93–118.

Cablk, M.E., and Minor, T.B. ,2003, Detecting and discriminating impervious cover with high-resolution IKONOS data using principal component analysis and morphological operators: International Journal of Remote Sensing, v. 24, no. 23, p. 4627–4645.

California Department of Forestry and Fire Protection, 2008, Wildland urban building codes: California Department of Forestry and Fire Protection Web site, accessed August 4, 2008, at http://www.fire.ca.gov/fire_prevention/fire_prevention_wildland_codes.php#code3.

California Tahoe Conservancy, 2005, Guidelines and criteria for the land acquisition program: California Tahoe Conservancy Web site, accessed September 16, 2009, at http://tahoe.ca.gov/files/2010_ACQ/2010_10_ESL_Acquisition_Criteria.pdf.

Candau, J., 2000, Calibrating a cellular automaton model of urban growth in a timely manner, in Parks, B.O., Clarke, K.M. , and Crane, M.P., eds., Proceedings of the 4th International Conference on Integrating Geographic Information Systems and Environmental Modeling—Problems, Prospects and Needs for Research, Sept. 2–8, 2000: University of Colorado, Boulder, CD-ROM.

Citygate Associates, LLC, 2004, Fire planning process for the urban-wildland interface in the City of South Lake Tahoe: Guidance document protection life, property, and community values through community-based planning: Citygate Associates, LLC, under contracted to City of South Lake Tahoe Fire Department.

Coats, R.N., and Goldman, C. R. 2001, Patterns of nitrogen transportation in streams of the Lake Tahoe Basin, California-Nevada: Water Resources Research, v. 37, no. 2, p. 405–415.

Coats, R., and Gunter, M., Heyvaert, A., Thomas, J., Luck, M. and Reuter, J., 2006, Water quality, watershed characteristics and land use in the Tahoe Basin: Journal of the Nevada Water Resources Association, v. 4, no. 1, p. 41–42.

Coats, R., Larsen, M., Heyvaert, A., James, T., Luck, M., and Reuter, J., 2008, Nutrient and sediment production, watershed characteristics, and land use in the Tahoe Basin, California-Nevada: Journal of the American Water Resources Association, v. 44, no. 3, p. 754–770.

Condon, P.M, Cavens, D., and Miller, N., 2009, Urban planning tools for climate change mitigation: Cambridge, Mass., Lincoln Institute of Land Policy, Policy Focus Report PF021, 52 p.

Conroy, M.J., Cohen, Y., James, F.C., Matsinos, Y.G., and Maurer, B.A., 1995, Parameter estimation, reliability and model improvement for spatially explicit models of animal populations: Ecological Applications, v. 5, no. 1, p. 17–19.

Cushman, S.A., and McGarigal, K., 2007, Multivariate landscape trajectory analysis—An example using simulation modeling of American marten habitat change under four timber harvest scenarios, in Bissonette, J.A., and Storch, I., eds., Temporal dimensions of landscape ecology: Springer, Wildlife Responses to Variable Resources, p. 119–140.

D'Erchia, F., Korschgen, C., Nyquist, M., Root, R., Sojda, R., and Stine, P., 2002, A framework for ecological decision support systems—Building the right systems and building the systems right: U.S. Geological Survey, Biological Resources Division, Information and Technology Report USGS/BRD/ITR-2001-0002, 50 p.

Dietzel, C., and Clarke, K., 2006, The effect of disaggregating land use categories in cellular automata during model calibration and forecasting: Computers, Environment, and Urban Systems, v. 30, p. 78–101.

Dobrowski, S.Z., Greenberg, J. A., Ramirez, C. M., and Ustin, S.L., 2006, Improving image derived vegetation maps with regression based distribution modeling: Ecological Modeling, v. 192, no. 1–2, p. 126–142.

Dramstad, W.E., Olson, J.D. , Forman, R.T.T., 1996, Landscape ecology principles in landscape architecture and land-use planning: Washington, D.C., Harvard University Graduate School of Design, Island Press, and the American Society of Landscape Architects, 77 p.

Duffie, E.S., Hessenflow, M., Wein, A., Halsing, D., Jeton, A., Schulz K., 2004. Tahoe Decision Support System—No-project alternative analysis: U.S. Geological Survey unpub. report submitted to the Tahoe Regional Planning Agency, 217 p.

Dunning, J.B., Jr., Stewart, D.J., Danielson, B.J., Noon, B.R., Root, T.L., Lamberson, R.H., Stevens, E.E., 1995, Spatially explicit population models—Current forms and future uses: Ecological Applications, v. 5, no. 1, p. 3–11.

Engelen, G., White, R., Uljee, I., Drazan, P., 1995, Using cellular automata for integrated modeling of socio-environmental systems: Environmental Monitoring and Assessment, v. 34, no. 2, p. 203–214.

Environmental Protection Agency, 2012, Watershed priorities—Lake Tahoe, CA & NV: Environmental Protection Agency Web site, accessed August 17, 2012, at available at http://www.epa.gov/region9/water/watershed/tahoe/.

Elliott-Fisk, D.L., and Erman, D.C., and Science Team Members, 1997, Sierra Nevada Ecosystem Project—final report to Congress plus Addendum (4 volumes): University of California, Centers for Water and Wildland Resources, Davis, 328 p.

Forney, W., Richards, L., Adams, K.D., Minor, T.B., Rowe, T.G., Smith, J.L., and Raumann, C.G., 2001, Land use change and effects on water quality and ecosystem health in the Lake Tahoe Basin, Nevada and California: U.S. Geological Survey Open-File Report 01–418, 29 p., available online at http://pubs.usgs.gov/of/2001/of01-418/.

Forney, W. M., Crescenti, N., and Oldham, I.B., 2008, Five methods of selecting targets for redevelopment: U.S. Geological Survey unpub. report submitted to the Tahoe Regional Planning Agency, 10 p.

Forney, W.M., and Oldham, I.B., 2011, The Lake Tahoe Basin Land Use Simulation Model: U.S. Geological Survey Open-File Report 2011–1275, 63 p., available at http://pubs.usgs.gov/of/2011/1275/.

Guzy, M.R., Smith, C.L., Bolte, J.P., Hulse, D.W., and Gregory, S. V., 2008, Policy research using agent-based modeling to assess future impacts of urban expansion into farmlands and forests: Ecology and Society, v. 13, no. 1, 37 p.

Halsing, D., Hessenflow, M., and Wein, A., 2005, The no-project alternative analysis—An early product of the Tahoe Decision Support System: Journal of the Nevada Water Resources Association, Lake Tahoe Edition, v. 2, no. 1, p. 15–28.

Halsing, D., 2006, Tahoe Land-Use Change Model summary report and climate change literature review and Tahoe Basin projections: U.S. Geological Survey unpub. report delivered to Lahontan Regional Water Quality Control Board and TetraTech, Inc., 14 p.

Hessenflow, M., and Halsing, D., 2006, A simulation model of land-use change in the Lake Tahoe Basin of California and Nevada, as used in a decision-support system, in Proceedings of the International Environmental Modelling and Software Society third biennial meeting—Summit on Environmental Modelling and Software, Burlington, Vermont, July 2006: International Environmental Modeling and Software Society Web site, accessed January, 2008, at http://www.iemss.org/iemss2006/papers/s8/S8_Hessenflow_Halsing.pdf.

Holt, R.D., Pacala, S.W., Smith, T.W., and Liu, J., 1995, Linking contemporary vegetation models with spatially explicit animal population models: Ecological Applications, v. 5, no. 1, p. 20–27.

Jankowski, P., Andrienko, N., and Andrienko, G., 2001, Map-centered exploratory approach to multiple criteria spatial decision making: International Journal of Geographical Information Science, v. 15, no. 2, p. 101–127.

Jassby, A.D., Reuter, J.E., Axler, R.P., Goldman, C.R., and Hackley, S.H., 1994, Atmospheric deposition of nitrogen and phosphorous in the annual nutrient load of Lake Tahoe (California-Nevada): Water Resources Research, v. 30, no. 7, p. 2207–2216.

Jenks, G.F, 1967, The data model concept in statistical mapping: International Yearbook of Cartography, v. 7, p. 186–190.

Kauffman, G.J., and Brant, T., 2000, The role of impervious cover as a watershed-based zoning tool to protect water quality in the Christina River Basin of Delaware, Pennsylvania, and Maryland, in Watershed Management Conference 2000: Water Environment Federation, p. 1–11.

Lahontan Regional Water Quality Control Board and Nevada Division of Environmental Protection, 2008 Lake Tahoe TMDL pollutant reduction opportunity report, v 2.0: Lahontan Regional Water Quality Control Board and Nevada Division of Environmental Protection, 279 p.

Lahontan Regional Water Quality Control Board and Nevada Division of Environmental Protection, 2009 Lake Tahoe total maximum daily load technical report: Lahontan Regional Water Quality Control Board and Nevada Division of Environmental Protection, 340 p., accessed July, 2010, at http://www.swrcb.ca.gov/rwqcb6/water_issues/programs/tmdl/lake_tahoe/docs/2_tmdl_techrpt.pdf.

Loftis, W.R., 2007, Soil survey of the Tahoe Basin area, California and Nevada: U.S. Department of Agriculture, Natural Resources Conservation Service, 2530 p., accessed November, 2009, at http://soildatamart nrcs.usda.gov/Manuscripts/CA693/0/Tahoe_CA.pdf.

Manley, P.N., Fites-Kaufman, J.A.A., Barbour, M.G., Schlesinger, M.D., and Rizzo, D.M., 2000, Lake Tahoe Watershed assessment, volume 1, chapter 5—Biological integrity: U.S. Forest Service, Pacific Southwest Research Station, General Technical Report PSW-GTR-175, p. 403–600.

Manley, P.N., Murphy, D.D., Schlesinger, M.D., Campbell, L.A., Merideth, M., Sanford, S., Heckmann, K., and Parks, S., 2007, The role of urban forests in conserving and restoring biological diversity in the Lake Tahoe Basin: Unpub. report submitted to the U.S. Forest Service Lake Tahoe Basin Management Unit in compliance with Southern Nevada Public Land Management Act Contract, 287 p.

McGarigal, K., and McComb W.C., 1995, Relationships between landscape structure and breeding birds in the Oregon Coast Range: Ecological Monographs, v. 65, no. 3, p. 235–260.

McGurk, B.J., and Davis, M.L., 1996, Camp and Clear Creaks, El Dorado County—Chronology and hydrologic effects of land use change, in Erman, D.C., ed., Status of the Sierra Nevada—Sierra Nevada Ecosystem Project Final Report to Congress: Davis, Calif., Centers for Water and Wildlands Research, v. 2, p. 1369–1406.

Merrill, A.G., 2001, Variation in the structure and nitrogen dynamics of mountain riparian zones: University of California Berkeley, unpub. Ph.D. dissertation, 251 p.

Minor, T., and Cablk, M., 2004, Estimation of impervious cover in the Lake Tahoe Basin using remote sensing and geographic information systems data integration: Journal of the Nevada Water Resources Association, v. 1, no. 1, p. 58–75.

National Interagency Fire Center, 2010, Fire information—Wildland fire statistics: National Interagency Fire Center Web site, accessed July 6, 2010, at http://www nifc.gov/fireInfo/fireInfo_statistics html.

National Science Foundation, 2009, Transitions and tipping points in complex environmental systems—A report by the NSF Advisory Committee for Environmental Research and Education: National Science Foundation, 56 p.

Nevada Department of Environmental Protection, 2012, Lake Tahoe watershed program: Nevada Department of Environmental Protection Web site, accessed August 17, 2012, at available at http://ndep nv.gov/bwqp/tahoe.htm.

Nevada Division of State Lands, 2008, Nevada Tahoe Resource Team: Nevada Division of State Lands Web site, accessed July 8, 2008, at available at http://lands nv.gov/program/tahoe.htm.

Pathway, 2008, Pathway 2007 Web page, accessed December 2008, at http://www.pathwaytahoe.org.

Peterson, M.N., Chen, X., and Liu J., 2008, Household location choices—Implications for biodiversity conservation: Conservation Biology, v. 22, no. 4, p. 912–921.

Raumann, C.G., and Cablk, M.E., 2008, Change in the forested and developed landscape of the Lake Tahoe Basin, California and Nevada, USA, 1940–2002: Forest Ecology and Management, v. 255, no. 8–9, p. 3424–3439.

Richards, L., 1999, An historical geographic information system for Truckee, California: California State University, Chico, unpub. Master's thesis, 135 p.

Riverson, J., Barreto, C., Shoemaker, L., Reuter, J., and Roberts, J., 2005, Development of the Lake Tahoe Watershed Model—Lessons learned through modeling in a subalpine environment, in Walton, R., ed., World Water Congress 2005—Impacts of global climate change: Proceedings of the World Water Congress, Anchorage, Alaska, May 15–19, 2005, 14 p.

Roberts, D.M., and Reuter, J.E., 2007, Draft Lake Tahoe total maximum daily load technical report, California and Nevada: California Regional Water Quality Control Board, Lahontan Region, and the Nevada Division of Environmental Protection, 341 p., accessed September 6, 2012, at http://www.swrcb.ca.gov/lahontan/water_issues/programs/tmdl/lake_tahoe/docs/laketahoe_tmdl_techrpt.pdf.

Rosenberg, B., and Guy, J., 1976, Prediction of beta from investment fundamentals: Financial Analysts Journal, v. 32, no. 3, p. 60–72.

Rowe, T.G., Saleh, D.K., Watkings, S.A., Kratzer, C.R., 2002, Streamflow and water-quality data for selected watershed in the Lake Tahoe Basin, California and Nevada, through 1998: U.S. Geological Survey Water Resources Investigations Report 02–4030, 125 p., available at http://pubs.usgs.gov/wri/wri024030/.

Simon, A., Langendoen, E.J., Bingner, R.L., Wells, R., Heins, A., Jokay, N., and Jaramillo, I., 2003, Lake Tahoe Basin framework implementation study—Sediment loadings and channel erosion: U.S. Department of Agriculture, Agricultural Research Service, National Sedimentation Laboratory Research Report No. 39. 377 p.

Steve Holl Consulting and Wildland Rx, 2007, Fuel reduction and forest restoration plan for the Lake Tahoe Basin wildland-urban interface: Steve Holl Consulting and Wildland Rx prepared for the Tahoe Regional Planning Agency, 15 p., accessed September 6, 2012, at http://www.trpa.org/documents/about_trpa/forest%20fules/Final/FFintro.pdf.

Stough, R.R., and Whittington, D., 1985. Multijurisdictional waterfront and land use modeling: Coastal Zone Management Journal, v. 13, no. 2, p. 151–175.

Swartzman, G.L., and Kaluzny, S.P., 1987, Simulation Model Evaluation, in Ecological simulation primer: New York, N.Y., MacMillan Publishing Company, p. 209–251.

Tahoe Environmental Research Center, 2009, Tahoe—State of the lake report: University of California Davis, Tahoe Environmental Research Center, accessed July 6, 2010, at http://terc.ucdavis.edu/stateofthelake/StateOfTheLake2009.pdf.

Tahoe Regional Planning Agency, 1996, 1996 evaluation—Environmental threshold carrying capacities and the regional plan package: Tahoe Regional Planning Agency.

Tahoe Regional Planning Agency, 2003, Best management practices retrofit program: Tahoe Regional Planning Agency.

Tahoe Regional Planning Agency, 2008a, Tahoe Regional Planning Agency Plan Area Statement maps: Tahoe Regional Planning Agency, accessed at various times in 2008 at http://www.trpa.org/default.aspx?tabid=204.

Tahoe Regional Planning Agency, 2008b, Discussion paper number 2—land use—PTOD, housing, and climate change issues relevant to policy changes being considered as regional plan amendments: Tahoe Regional Planning Agency memorandum, March 24, 2008, p. 21–25, accessed August 2008, at http://www.trpa.org/documents/packets/apc_packets/2008_apc_packets/may_2008_apc_packet.pdf.

Tahoe Regional Planning Agency, 2008c, Tahoe Regional Planning Agency: Tahoe Regional Planning Agency Web page, accessed various times from January to December 2008, at http://www.trpa.org.

Tahoe Regional Planning Agency, 2008d, Transect zoning map—alternative 2: Tahoe Regional Planning Agency Web page, accessed various times in fall 2008, http://www.trpa.org/documents/rp_update/Transect_Map_12-12-08.pdf.

Tetra Tech, Inc., 2007, Watershed hydrologic modeling and sediment and nutrient loading estimation for the Lake Tahoe total maximum daily load—Final modeling report: Tetra Tech, Inc., prepared for the Lahontan Regional Water Quality Control Board and the University of California, Davis, 118 p., accessed September 6, 2012, at http://www.waterboards.ca.gov/rwqcb6/water_issues/programs/tmdl/lake_tahoe/docs/peer_review/tetra2007.pdf.

Turner, M.G., Gardner, R.H., and O'Neill, R.V., 2001, Landscape ecology in theory and practice—Pattern and process: New York, N.Y., Springer-Verlag, 401 p.

U.S. Forest Service, 2008, Urban lot management program—Land acquisitions and the Santini-Burton Act: U.S. Forest Service, Lake Tahoe Basin Management Unit, Web site, accessed July 8, 2008, at http://www r5.fs fed.us/ltbmu/forest-management/urban-lots/santini_burton.shtml.

Urban, D.L, 2000, Using model analysis to design monitoring programs for landscape management and impact assessment: Ecological Applications, v. 10, no. 6, p. 1820–1832.

Veith, T.L., Wolfe, M.L., Heatwole, C.D., 2003, Optimization procedure for cost effective BMP placement at a watershed scale: Journal of the American Water Resources Association, v. 39, no. 6, p. 1331–1343.

Verburg, P.H., Soepboer, W., Veldkamp, A., Limpiada, R., Espaldon, V., Mastura, S.S.A., 2002, Modeling the spatial dynamics of regional land use—The Clue-S Model: Environmental Management, v. 30, no. 3, p. 391–405.

Weller, D. E., Jordan, T.E., and Correll, D.L., 1997, Heuristic models for material discharge from landscapes with riparian buffers: Ecological Applications, v. 8, no. 4, p. 1156–1169.

With, K.A. 2007, Invoking the ghosts of landscapes past to understand the landscape ecology of the present and the future, in Bissonette, J. A., and Storch, I., eds.,Temporal dimensions of landscape ecology—Wildlife responses to variable resources: New York, N.Y., Springer Science+Business Media, LLC, p. 43–58.

Wu, F., 1998, Simulating urban encroachment on rural land with fuzzy-logic controlled cellular automata in a geographical information system: Journal of Environmental Management, v. 53, p. 293–308.

Yearley, S., 1999, Computer models and the public's understanding of science—A case study analysis: Social Studies of Science, v. 29, no. 6, p. 845–866.